Studies in History and Philosophy of Science

Volume 45

More information about this series at http://www.springer.com/series/5671

Lesley B. Cormack • Steven A. Walton
John A. Schuster
Editors

Mathematical Practitioners and the Transformation of Natural Knowledge in Early Modern Europe

Springer

Editors
Lesley B. Cormack
Department of History and Classics
University of Alberta
Edmonton, AB, Canada

Steven A. Walton
Department of Social Sciences
Michigan Technological University
Houghton, MI, USA

John A. Schuster
Unit for History and Philosophy of Science
University of Sydney
Sydney, Australia

ISSN 0929-6425 ISSN 2215-1958 (electronic)
Studies in History and Philosophy of Science
ISBN 978-3-319-49429-6 ISBN 978-3-319-49430-2 (eBook)
DOI 10.1007/978-3-319-49430-2

Library of Congress Control Number: 2017933848

Printed on acid-free paper

This Springer imprint is published by Springer Nature
The registered company is Springer International Publishing AG
The registered company address is: Gewerbestrasse 11, 6330 Cham, Switzerland

Contents

About the Editors and Authors

Lesley B. Cormack is a historian of science, now dean of arts at the University of Alberta. She is the author of *Charting an Empire: Geography at the English Universities 1580–1620* (Chicago, 1997) and *A History of Science in Society: From Philosophy to Utility* with Andrew Ede (Broadview Press, 2004, 2nd Edition; University of Toronto Press 2012, 3rd Edition, 2016) and editor of *Making Contact: Maps, Identity, and Travel* (University of Alberta Press, 2003) and *A History of Science in Society: A Reader* (Broadview Press, 2007). She is now completing a book on the development and use of the Molyneux globes in sixteenth-century England.

Fokko Jan Dijksterhuis is associate professor in history of science at the University of Twente and extraordinary professor in early modern knowledge history at Free University, Amsterdam. He took a degree in mathematics and science studies, obtaining his doctorate with a study of Christiaan Huygens and seventeenth-century optics, *Lenses and Waves*. He is interested in early modern knowledge practices, in particular exact ways of knowing and doing and everything related to light, color, and vision. The contribution to this volume was part of the research project "The Uses of Mathematics in the Dutch Republic," funded by the Netherlands Organization for Scientific Research (Nederlandse Organisatie voor Wetenschappelijk Onderzoek, or NWO), on the cultural history of early modern mathematization.

Sven Dupré is professor of history of art, science, and technology at Utrecht University and the University of Amsterdam. He is the PI of the "Technique in the Arts: Concepts, Practices, Expertise, 1500–1950" (ARTECHNE) project that is supported by a European Research Council (ERC) Consolidator Grant. Previously, he was professor of history of knowledge at the Freie Universitat and director of the research group "Art and Knowledge in Premodern Europe" at the Max Planck Institute for the History of Science in Berlin. He is also actively involved in research in technical art history at the Atelier Building in Amsterdam, where the Rijksmuseum, the Cultural Heritage Agency of the Netherlands, and the University of Amsterdam combine their knowledge in the field of restoration and preservation of art objects. He has published on a wide range of topics in the history of early modern science, technology and art. Recent publications include *Early Modern Color Worlds* (Brill, 2015), *Embattled Territory: The Circulation of Knowledge in the Spanish Netherlands* (Academia Press/LannooCampus, 2015), *Laboratories of Art: Alchemy and Art Technology from Antiquity to the 18th Century* (Springer, 2014), *Art and Alchemy: The Mystery of Transformation* (Hirmer, 2014) in conjunction with an exhibition at the Museum Kunstpalast in Düsseldorf, and *Translating Knowledge in the Early Modern Low Countries* (LIT, 2012).

Alex G. Keller studied history at Cambridge and Oxford, completing a PhD at Cambridge on *Early Printed Books of Mechanical Inventions 1569–1629*. He published an anthology of pictures from these books, as *A Theatre of Machines* (1964). For many years, he taught history of science at the University of Leicester, where he is now an honorary fellow in the School of Historical Studies. He has taught at Case Western Reserve University in Cleveland, Ohio, and has been a research fellow at the Smithsonian Institution. He has published extensively on Renaissance mechanics and engineering, culminating in a translation from Spanish, with extensive commentary, of a late sixteenth-century five-volume manuscript technical encyclopedia, *Los Veintiun Libros de los Ingenios y de las Maquinas*, as *The Twenty-One Books of Engineering and Machines* (1996). Keller has also ventured into the early twentieth-century teaching and writing (*The Infancy of Atomic Physics*) and visited as a lecturer in Spain (Zaragoza, Santander) and Sweden. Keller was the editor of *Icon*, the journal of ICOHTEC (the International Committee for the History of Technology) until 2009.

W.R. Laird took his PhD in medieval studies from the University of Toronto, with a dissertation on the *scientiae mediae* in the Middle Ages. He taught in the Department of History, Rice University, and in the Institute for the History and Philosophy of Science and Technology, University of Toronto, before settling at Carleton University in Ottawa, where he teaches ancient and medieval intellectual history and the history of science in the College of the Humanities and in the Department of History. He is the author of *The Unfinished Mechanics of Giuseppe Moletti*, which is an edition, translation, and study of a sixteenth-century mechanical treatise, and of a number of articles on medieval, Renaissance, and early modern science, with a special emphasis on mechanics and the science of motion. He is currently writing a history of mechanics in the sixteenth century, to be called *The Renaissance of Mechanics*.

John A. Schuster is honorary research fellow in the Unit for History and Philosophy of Science and Sydney Centre for the Foundations of Science, University of Sydney, and honorary fellow, Campion College, Sydney, the only private liberal arts college in Australia. He previously taught at Princeton, Leeds, Cambridge, and the University of New South Wales. He has published on the historiography of the scientific revolution, the nature and dynamics of the field of early modern natural philosophy, Descartes' natural philosophical and mathematical career, the problem of the origin of experimental sciences in the seventeenth and eighteenth centuries, and the political and rhetorical roles of scientific method. Recent publications include *Descartes-Agonistes: Physico-Mathematics, Method and Corpuscular-Mechanism—1618-33* (Springer, 2013) and "Cartesian Physics" in *The Oxford Handbook of the History of Physics* (2013): 56–95.

Steven A. Walton teaches history of science and technology, European history, and military history at Michigan Technological University, where he is also actively involved with the graduate program in industrial heritage and archaeology. His primary scholarly writing is on the intersections between science, technology, and the military, particularly in the early modern European and antebellum American world. He has just published the travel diaries of Thomas Kelah Wharton, a nineteenth-century architect and artist, and an article on US Civil War artillery and is working on a book on *Transitions in Defense*, on changes in fortification practice and rationale in sixteenth-century England. He has edited works on *Fifty Years of Medieval Technology and Social Change* (Ashgate, forthcoming), *Wind & Water in the Middle Ages: Fluid Technologies from Antiquity to the Renaissance* (ACMRS, 2006), and *Instrumental in War: Science, Research, and Instruments Between Knowledge and the World* (Brill, 2005).

List of Figures

Chapter 1
Introduction: Practical Mathematics, Practical Mathematicians, and the Case for Transforming the Study of Nature

Lesley B. Cormack

Abstract This book argues that we can only understand the transformations of nature studies in the early modern period, often called the Scientific Revolution, if we take seriously the interaction between those who know by doing (practitioners or craftsmen) and those who know by thinking (scholars or philosophers). Mathematical practitioners played an essential role in this transformation; this book examines the role of mathematics and mathematical practice on the changing ideology and methodology of science. We first set out the problematic, examining the argument from both sides: articulating Zilsel, Cormack identifies those dimensions of practical mathematics that showed up as important aspects of 'the new science'; Schuster focuses on the new scientists as selective appropriators of ideas, values and practices originally embedded in practical mathematics. This book furthers the debate about the role of mathematical practice in the scientific revolution in four ways. First, it demonstrates the variability of practical mathematicians and of their practices. Second, it argues that in spite of this variability, participants were able to recognize the family resemblance between the different types. Third, differences and nuances in practical mathematics typically depended on the different contexts in which it was practiced. Fourth, this book shows that diverse and new historiographical approaches to the study of practical mathematics should be considered.

Theory and practice; scholar and craftsman. Historical discussions of the investigation of nature have often been seen through the lens of such dichotomies, particularly those concerning the early modern period. This book takes the position that we can only understand transformations of nature studies in the early modern period, often called the Scientific Revolution, if we take seriously the interaction between those who know by doing (practitioners or craftsmen) and those who

L.B. Cormack (✉)
Department of History and Classics, University of Alberta, Edmonton, AB, Canada
e-mail: lcormack@ualberta.ca

L.B. Cormack et al. (eds.), *Mathematical Practitioners and the Transformation of Natural Knowledge in Early Modern Europe*, Studies in History and Philosophy of Science 45, DOI 10.1007/978-3-319-49430-2_1

know by thinking (scholars or philosophers). These are not in opposition however. Rather, theory and practice are end points on a continuum, with some practitioners interested only in the practical, others only in the theoretical, and most inhabiting and moving through the murky intellectual world in between. It is this liminal space, this trading zone or borderland, where influence, appropriation, and collaboration could lead to new methods, new subjects of enquiry, and new social structures of natural philosophy and science.

Understanding that this continuum exists leads to new and important insights into what happened in the period of the Scientific Revolution. Historians have long seen the sixteenth and seventeenth centuries as fundamentally important to an understanding of the changing study of nature. An earlier generation of scholars argued that a philosophical and theoretical change 'from a closed world to an infinite universe'[1] was the key and that lowly practitioners were in no way connected with that change. However, this period in European history was also a period of great technological, economic, and social changes, and it has become increasingly more difficult for historians to maintain that these changes were in no way related to conceptual shifts in the understanding of nature or *vice versa*. Historians coming to maturity during the 1980s and 1990s were struck by the need for a more contextualized, more complicated interpretation of the Scientific Revolution, and have begun to examine the interconnections between scientific understanding and practice, and even to argue that practice and practical knowledge were necessary components to the changes taking place in natural philosophy and its methodologies.

The case for influence, collaboration or appropriation between theory and practice can be most persuasively drawn in the area of mathematics, since practical mathematics was a growing field in early modern Europe. The 'mixed mathematics' of Aristotle, transforming into the more capacious category of practical mathematics by the sixteenth century, had a long history of investigation by both scholars and craftsmen. During the sixteenth and seventeenth centuries, more and more men began to use mathematics in order to measure and control their environment, in areas such as surveying, navigation, military arts, and cartography. Given that one of the questions for historians and philosophers of the Scientific Revolution is how the mathematization of natural philosophy came into being, an investigation of the interplay between useful mathematics and its practitioners on the one hand, and natural philosophers on the other, seems in order.[2] This book provides an important step in the examination of this relationship.

[1]Alexandre Koyré, *From a Closed World to an Infinite Universe* (Baltimore: Johns Hopkins University Press, 1957).

[2]Geoffrey Gorham and Benjamin Hill (eds.), *The Language of Nature: Reassessing the Mathematization of Natural Philosophy* (Minneapolis: University of Minnesota Press, 2015) provides an interesting philosophical discussion of this issue. See also Lesley B. Cormack, "*The Grounde of Artes*: Robert Recorde and the Role of the Muscovy Company in the English Mathematical Renaissance," *Proceedings of the Canadian Society for the History and Philosophy of Mathematics* 16 (2003): 132–138.

1.1 E.G.R. Taylor and Mathematical Practitioners

In 1954, E.G.R. Taylor, a well-respected historian of geography, turned her attention to what she termed 'mathematical practitioners', men who deployed mathematical concepts for practical ends. She gathered a collective biography of almost 600 individuals practicing from 1485 to 1714.[3] Taylor defined mathematical practitioners as men who earned their living by teaching, writing, constructing and selling instruments, and acting in technical capacities. She argued that they should be considered essential players in the evolution of natural knowledge, although her goal was primarily retrieving their histories rather than making larger theoretical and historiographical claims.

Taylor demonstrated that, beginning in the sixteenth century, a number of both university-trained and self-taught men set themselves up as mathematics teachers and practitioners. These men sold their expertise as teachers through publishing textbooks, making instruments, and offering individual and small group tutoring. In the process, they argued for the necessity of practical knowledge of measurement, winds, surveying, artillery, fortification, and mapping, rather than for a more philosophical and all-encompassing knowledge of the natural world.

Most mathematical practitioners were university-trained, showing that the separation of academic and entrepreneurial teaching was one of venue and emphasis, rather than necessarily one of background. Mathematical practitioners claimed the utility of their knowledge, a rhetorical move that encouraged those seeking such information to regard it as useful.[4] It is impossible to know the complete audience for such expertise, but English mathematical practitioners, at any rate, seem to have aimed their books and lectures at an audience of London gentry, merchants, and occasionally artisans.[5] It is probably this choice of audience that most influenced their emphasis on utility, since London gentry and merchants were looking for practicality and means to improve themselves and their businesses. At the same time, the expanding English state looked favourably on practical schemes that could facilitate this expansion.

Mathematical practitioners professed their expertise in a variety of areas, especially such mathematical applications as navigation, surveying, gunnery, and forti-

---

[3]E.G.R. Taylor, *The Mathematical Practitioners of Tudor and Stuart England* (Cambridge, Cambridge University Press, 1954). Taylor completed a second volume, *Mathematical Practitioners of Hanoverian England, 1714–1840* (Cambridge: Cambridge University Press, 1966), published just after her death, and bringing the number of mathematical practitioners she identified to 2500.

[4]Katherine Neal, "The Rhetoric of Utility: Avoiding Occult Associations for Mathematics through Profitability and Pleasure," *History of Science* 37 (1999): 151–178 discusses some attempts to make mathematics appear useful.

[5]Thomas Hood's lecture, *A Copie of the Speache made by the Mathematicall Lecturer, unto the Worshipfull Companye present . . . in Gracious Street: the 4 of November 1588* (n.p. 1588) is a good example. See Deborah Harkness, *The Jewel House. Elizabethan London and the Scientific Revolution* (New Haven: Yale University Press, 2007) for a discussion of the complex interactions among London merchants, artisans and scholars.

fication. For example, Galileo's early works on projectile motion and his innovative work with the telescope were successful attempts to gain patronage in the mathematical realm.[6] Simon Stevin claimed the status of a mathematical practitioner, including an expertise in navigation, fortification, and surveying. William Gilbert argued that his larger natural philosophical arguments about the magnetic composition of the earth had practical applications for navigation.

1.2 Taylor's Category Continues

Other historians of science have taken up the challenge of understanding the role of mathematics and its practitioners in changes to the scientific landscape of the sixteenth and seventeenth centuries. While Mario Biagioli demonstrated that mathematics had less status than philosophy in early modern Italian universities, arguing that this accounted for Galileo's move from professor of mathematics to that of philosophy and then on to Court philosopher, others have argued that mathematics was an important component of changing methods and theories of natural philosophers.[7] Taylor's middle category of mathematical practitioner who did not need an explicit connection to natural philosophy, or natural philosophers for that matter, merely punted on this question. Both Jim Bennett and Stephen Johnston have taken Taylor's category very seriously. Johnston has looked at the métier of these artisans in a variety of venues, including earthworks, engineering, and instrument making.[8] Bennett has been particularly interested in instrument-makers and their activities, and in the process has shown that the move to a mechanical philosophy on the part of high-status natural philosophers owes much to the mechanics' art.[9] Peter Dear examined Jesuit mathematicians to see whether their school of mathematical physics provided a key connection between theory

[6]Of course, once Galileo successfully gained a patronage position, particularly with the Florentine Medici court, he left his mathematical practitioner roots behind and became a much higher status natural philosopher. Mario Biagioli, *Galileo, Courtier: The Practice of Science in the Culture of Absolutism.* (Chicago: University of Chicago Press, 1993). Matteo Valeriani, *Galileo Engineer* (Dordrecht: Springer, 2010).

[7]Mario Biagioli, "The Social Status of Italian Mathematicians, 1450–1600," *History of Science* 27 (1989): 41–95, and his *Galileo Courtier* (n.6, above).

[8]Stephen Johnston, *Making Mathematical Practice: Gentlemen, practitioners and artisans in Elizabethan England*, PhD dissertation, Cambridge University, 1994, and "Mathematical Practitioners and Instruments in Elizabethan England," *Annals of Science* 48.4 (1991): 319–44.

[9]James A. Bennett, "The Mechanics' Philosophy and the Mechanical Philosophy," *History of Science* 24 (1986): 1–28, and "The Challenge of Practical Mathematics," pp. 176–190 in S. Pumfrey, P.L. Rossi and M. Slawinski (eds.), *Science, Culture and Popular Belief in Renaissance Europe* (Manchester: Manchester University Press, 1991).

and practice.[10] He argued that mathematics allowed these Catholic philosophers to avoid some of the more inflammatory parts of the new philosophy, and to create a new 'physico-mathematics'. Approaching the issue from a different angle, Eric Ash examines the Dover Harbor engineering project, seeing the interaction of these practitioners and their development of 'expertise' (rather than philosophy) as an important status marker.[11]

1.3 Framing the Argument

This book is concerned with the role of these mathematical practitioners in changes to the study of nature in the sixteenth and seventeenth centuries. We have placed the practitioner in the centre of the story, examining both the practical and philosophical implications of his participation in the continuum of nature studies. In doing so, we hope to place a mirror of sorts between theory and practice and use the practitioner to gaze in both directions.[12] We begin in the introductory section by setting out the problematic, both in terms of the older 'Zilsel' thesis, which argued that skilled artisans and mathematical practitioners were essential for the transformation of natural knowledge known as the Scientific Revolution, and through an exploration of how it might be possible for practitioners and natural philosophers to have interacted and in what ways that might have happened. Through two opening chapters with differing explanatory models, we present a two-sided problematic through which to read the case studies that follow in the two subsequent sections. In Chap. 2, "Handwork and Brainwork: Beyond the Zilsel Thesis", Lesley Cormack provides the historiographical framing for this discussion: Edgar Zilsel developed the best early example of the question of the role of mathematical practitioners (or as he would have called them, "superior artisans") in the Scientific Revolution, and so we begin with his thesis. Cormack thus presents the case for the importance of social, economic, and cultural influences on the changing face of nature studies, particularly seeing the importance of mathematical practitioners in putting forward an agenda of utility, measurement, and inductive methodology. This is an argument for the important influence of both social factors and the practitioners themselves.

On the other hand, John Schuster, in Chap. 3, "Consuming and Appropriating Practical Mathematics and the Mixed Mathematical Fields", argues that if math-

[10]Peter Dear, *Discipline and Experience: The Mathematical Way in the Scientific Revolution* (Chicago: University of Chicago Press, 1995). See Schuster, Chap. 3, for a critique of Dear's argument.

[11]Eric Ash (ed.), *Expertise: Practical Knowledge and the Early Modern State*, *Osiris* 25 (2010); *Power, Knowledge and Expertise in Elizabethan England* (Baltimore: Johns Hopkins University Press, 2004).

[12]In certain ways, this approach was long ago championed by Edwin T. Layton, "Mirror-Image Twins: The Communities of Science and Technology in Nineteenth-Century America," *Technology & Culture* 12 (1971): 562–580.

ematical practice and practitioners were relevant to the study of nature in the early modern period, it is important historiographically to clarify this relation. He criticizes historical narratives which speak of practical or mixed mathematics 'influencing' or 'shaping' natural philosophy and proposes that the relationship is better understood as a process of appropriating and translating resources from one field to the other. He also questions explanations in which some aspect of practical mathematics directly causes a correspondingly essential change in the study of nature. Thus, in this introductory section, the argument is examined from both sides: articulating Zilsel, Cormack identifies those dimensions of practical mathematics that showed up as important aspects of 'the new science', while Schuster focuses on the new scientists as selective appropriators of ideas, values and practices originally embedded in practical mathematics.

1.4 Structure of the Volume

The contributors then take on two distinct parts of this argument. In Part 1, "What Did Practical Mathematics Look Like?", we investigate the state of mathematical practice in a number of European countries, especially England, the Dutch Republic, Italy and France. Just what was practical mathematics? Is this term more properly used to describe the 'seat-of-the-pants' calculations of gunners, as Steve Walton might argue? Or was mathematical practice the work of mathematical instrument-makers and instructors, who were better educated and mingled with the gentry and virtuosi, as Lesley Cormack suggests? What role did material artifacts, such as instruments and machines (the latter discussed by Alex Keller), have in changing thinking about nature?

In Chap. 4, "Mathematics for Sale", Cormack investigates the location of mathematics within London. She examines mathematical lectures and especially instrument-makers both inside and outside the City walls. Cormack discovers a vibrant practical mathematical community, whose members were gentry, scholars, merchants, instrument-makers, and navigators. She does not find, however, that these men or their ideas changed natural philosophy in a direct way.

Steven Walton, in Chap. 5, "Technologies of Pow(d)er", investigates the life and work of Edmund Parker, a gunner for Queen Elizabeth, in order to examine the role of mathematics in the very practical world of artillery. What he discovers is that mathematics was more useful as a social object, helping its practitioners to gain status, than as a tool to develop new understandings or even better practices in the area of artillery and fortification.

Alex Keller, in Chap. 6, "Machines as Mathematical Instruments", examines how Leonardo's prescient conjoining of the technological realm of machinery and the philosophical field of mechanics—which in his lifetime had little if anything to actually do with one another—became a reality (or at least was thoroughly believed if not proven) by the end of the sixteenth century. He locates the forces that united the two realms in Renaissance commentaries on the pseudo-Aristotelian *Mechanical*

Problems, the impulse towards invention that appealed to geometry and mathematics (the so-called "Theatre of Machines" tradition), and from the rise of both utility and philosophical sophistication of scientific instruments.

In Part II, "What was the Relationship between Practical Mathematics and Natural Philosophy?", we seek to understand the relationship between natural philosophy and practical mathematics, in all its particularities, this time considering it from the vantage point of natural philosophy. Sven Dupré, in Chap. 7, "The Making of Practical Optics", suggests that mathematical practitioners were not all alike, and so, while there is definitely a connection between practice and theory, the connection is likely different for each type of practitioner. Dupré argues that opticians appropriated perspective traditions in order to create practical optics (harkening back to Schuster's argument in Chap. 3), but shows just how complicated this story must be, given that each practitioner uses different theories to different ends.

In Chap. 8, "Hero of Alexandria and Renaissance Mechanics", W.R. Laird shows us an instance in which mixed mathematics was not appropriated for natural philosophizing, examining the Hero of Alexandra tradition of pneumatics. This practical (or at least amusing) study of automata and other devices was a dead end for natural philosophy, arguing against a connection between at least one branch of practical mathematics and changes to natural philosophy. On the other hand, Giuseppe Moletti, a natural philosopher, was certainly interested in mathematical machines, at least as oddities. So mixed mathematics certainly did draw the attention of natural philosophers.

Fokko Dijksterhuis, in Chap. 9, "*Duytsche Mathematique* and the Building of a New Society", examines the transformation from practical to theoretical mathematics as a move to increase the status of mathematics, both for natural philosophy and for the mathematicians themselves. By examining the different ways that practical mathematics was introduced into the educational and court systems in the two states of Friesland and Holland, Dijksterhuis shows that this was a complex and deeply contingent development.

1.5 Conclusion

From these case studies it becomes apparent that, just as there were many types of mathematical practice and practitioners, there were individualized connections and interactions between practice and theory. Some practitioners and practice did not influence natural philosophy, but others did, and the mathematization of nature developed, and with it, a sense of the utility of mathematical and natural philosophical knowledge. Clearly mathematics and mathematical practice were important to the fields of navigation, ballistics, surveying, instrument-making, and all the cognate fields of mixed and practical mathematics. Admittedly, why they were intellectually or socially important and how functionally necessary they were varied from case to case, but in every case practitioners and philosophers alike appealed to mathematics to support their positions. Equally, mathematics and mathematical practice were important to the changing attitudes towards measurement as knowledge, towards the role of exactitude and uncertainty in truth-claims, and thus to the transformation of natural knowledge in the early modern period.

This book therefore furthers the debate about the role of mathematical practice in the scientific revolution in four ways. First, these essays demonstrate the variability of the identity of practical mathematicians and of the practices involved in their activities in early modern Europe. This decommissions simplistic old questions such as 'did practical mathematics shape the new science of the Scientific Revolution?' We know that it did, but it did so differently in the various mathematical and philosophical sub-disciplines. Thus the argument must be more nuanced, taking into account the multiplicity of mathematic practices in order to have a multi-faceted and nuanced answer. The answer is at once 'yes' and 'it depends' and 'in some ways and not others'. In other words, these essays demand that we re-examine the overarching narrative about the interaction of practical mathematics and natural philosophy, while insisting that the full continuum of practice and understanding be taken into account to understand this era.

Second, although practical mathematical knowledge was transmitted and circulated in a wide variety of ways in early modern Europe, participants were able to recognize the family resemblance between the different types. This kinship allowed practitioners and scholars to see connections and contrasts. This makes it quite reasonable to say that despite diversity, practical mathematics did constitute a culture or a definable community, and as Deborah Harkness has shown, these practitioners circulated amongst each other as a sort of extended family.[13]

Third, differences and nuances in practical mathematics typically depended on the different contexts in which it was practiced. Social, cultural, political, and economic particularities do matter. Thus, the identities of mathematical practitioners, the methods they used, their putative and real audiences, and the media employed could all vary, even while the mathematical public or community could see connections.

Fourth and finally, this book shows that diverse and new historiographical approaches to the study of practical mathematics should be considered. In order to understand the interaction between theory and practice, scholar and craftsman, practical mathematician and natural philosopher, we will need to use different approaches. We will need to examine the historiographical tension between appropriation and influence, attention and efficacy. We will need to re-examine the relations amongst disciplines, and to take seriously to differences and connections among practical mathematics, mixed mathematics, and natural philosophy.

[13]Harkness, *The Jewel House* (n.5, above).

Part I
Framing the Argument: Theories of Connection

Chapter 2
Handwork and Brainwork: Beyond the Zilsel Thesis

Lesley B. Cormack

Abstract This chapter challenges the traditional historiography of the scientific revolution, arguing that skilled artisans and mathematical practitioners were essential for a transformation of natural knowledge, the so-called 'scholar-craftsman' debate. Beginning with a new articulation of Edgar Zilsel's thesis, which argued for an essential role for mathematical practitioners (or as he would have called them, "superior artisans") in the scientific revolution, this chapter argues that historians need to take into account social, cultural, political and economic factors, rather than the simpler Marxist explanations of Zilsel. Cormack thus presents the case for the importance of social, economic, and cultural influences on the changing face of nature studies, particularly seeing the importance of mathematical practitioners in putting forward an agenda of utility, measurement, and inductive methodology. This is an argument for the important influence of both social factors and the practitioners themselves. Using English geography in the sixteenth century, and particularly the work of Edward Wright and Thomas Harriot, she argues that geography and mathematics allowed communication between theory and practice, provided new spaces for such exchanges, and changed attitudes towards mathematization, practicality and utility.

2.1 Introduction

The scientific revolution has long been a central explanatory concept in the history of science.[1] Since the seventeenth century, analysts of this change into modernity have argued for the fundamental importance of the sixteenth and seventeenth

[1]For instance, Thomas Kuhn, *Structure of Scientific Revolutions* (Chicago: University of Chicago Press, 1962), uses the scientific revolution as its central example.

L.B. Cormack (✉)
Department of History and Classics, University of Alberta, Edmonton, AB, Canada
e-mail: lcormack@ualberta.ca

L.B. Cormack et al. (eds.), *Mathematical Practitioners and the Transformation of Natural Knowledge in Early Modern Europe*, Studies in History and Philosophy of Science 45, DOI 10.1007/978-3-319-49430-2_2

centuries in creating a new construction of the world.[2] Indeed, the twentieth-century discipline of history of science really began by focusing on the problem of the origin of modern science, and the work of some of its great founders concentrated on what this important transformation was and how it came to take place.[3]In more recent years, however, the term and the coherence of the ideas and events encompassed within it have been brought under considerable scrutiny. By 1988, the debate had gone so far that at the first Anglo-American History of Science meeting in Manchester, Jan Golinski could ask "the question as to whether the notion of a coherent, European-wide, Scientific Revolution can survive continued historiographical scrutiny".[4] In 1996, Steven Shapin began his analysis of the period with the now oft-quoted phrase, "There is no such thing as the scientific revolution and this is a book about it".[5] Although the term continues to be used, and, the conservatism of university curricula being what it is, will continue to stand as the title of numerous courses for many years to come, are Golinski and Shapin right? Has the term 'scientific revolution' outlived its usefulness?

In the 1930s, when Edgar Zilsel began working on his project concerning the origins of modern science, the belief that modern science had begun in the early modern period was well established.[6] Burtt had already published his famous book, *The Metaphysical Foundations of Modern Science* (1924), establishing a philosophical change in worldview as the foundational moment for modern science.[7] Most postwar, philosophically-minded historians of science followed suit, creating an explanatory structure largely unquestioned until the 1980s. Although Zilsel's Marxist and socological background placed him in an opposing camp, he took for granted the existence of modern science after 1700 and the reality of its evolution in the preceding centuries. For Zilsel, science, once achieved, would not be subject to material pressures, since it would represent the truth. What had to be explained

[2]See David Lindberg's "Introduction," in *Reappraisals of the Scientific Revolution*, Lindberg and Robert Westman, eds. (Cambridge: Cambridge University Press, 1990), 1–26, for a historical appraisal of the early use of this term.

[3]Edwin A. Burtt, *The Metaphysical Foundations of Modern Science* (London: K. Paul, 1924); Herbert Butterfield, *The Origins of Modern Science, 1300–1800* (London: Bell, 1949); and Alexandre Koyré, *From a Closed World to an Infinite Universe* (Baltimore: Johns Hopkins University Press, 1957), most particularly. Robert K. Merton, "Science, Technology and Society in Seventeenth-Century England," *Osiris* 4 (1938); second edition, (New York: Harper and Row, 1970) employs a different type of analysis, but has a similar definition of the scientific revolution, as does J. Dijksterhuis, *The Mechanization of the World Picture* (Oxford: Oxford University Press, 1961).

[4]Jan Golinski, Introduction to "The Scientific Revolution in its Social Context," session at the BSHS and HSS Joint Conference, Manchester, July 11–15, 1988, Abstracts, 1.

[5]Steven Shapin, *The Scientific Revolution* (Chicago: University of Chicago Press, 1996), 1.

[6]See Deiderick Raven and Wolfgang Krohn, "Introduction," *Edgar Zilsel: The Social Origins of Modern Science*, Deiderick Raven, Wolfgang Krohn and Robert S. Cohen, eds., (Dordrecht, Boston: Kluwer Academic Publishers, 2000) for an appraisal of the intellectual climate in which Zilsel worked.

[7]See Lindberg, 16, and F. Cohen, 88–97, for a fuller treatment of Burtt.

were the sociological preconditions that would allow the truth to emerge. Thus, from completely different political and epistemic points of view, Zilsel and the great triumverate of Burtt, Butterfield and Koyré came to similar definitions of what was soon called the 'scientific revolution' and to a similar chronological moment for its emergence.

In recent years, however, historians, and especially sociologists of knowledge, have become less convinced that some monolith called 'science' was discovered in this period. They have also questioned the revolutionary nature of the changes to the early modern investigation of the natural world. Thus, both parts of the term – 'scientific' and 'revolution' – have been challenged. The revolutionary nature of the scientific change in this period was the first to be questioned. Medievalists such as Pierre Duhem argued for continuity with an earlier period, thereby denying the revolutionary nature of the sixteenth or seventeenth centuries. Others questioned whether changing ideas about, for example, the ordering of the universe, affecting only a few hundred people at most and taking over 150 years to convince even those, could be called revolutionary.[8] Even for Butterfield, the lag of chemistry and biology was a serious issue, causing him to claim that the scientific revolution took 500 years. Furthermore, the more fundamental issue of whether the topics investigated were even science has now come to the forefront. Most historians of this period would now cautiously use the term 'natural philosophy' rather than 'science' when dealing with the early modern period.[9] But most continue to look for something identifiable as the origins of modern science. Cunningham and Williams challenged that assumption. As they have so provocatively pointed out, 'natural philosophy' was not simply another word for 'science' but referred to an essentially theological and philosophical investigation of the natural world. Those embarked on this enterprise were not scientists but natural philosophers.[10] Thus, according to Cunningham and Williams, this was not a revolution into science, but if anything a philosophical revolution. If this event, the 'scientific revolution', was neither scientific, nor revolutionary, does anything remain?

[8]Pierre Duhem, *Les Origines de la Statique*, 2 vols. (Paris, 1905–6); Lynn Thorndike, *History of Magic and Experimental Science*, 8 vols. (New York: Macmillan, 1923–58), for example. See Lindberg, 13–15. Paul K. Feyerabend, *Against Method* (London: New Left Books, 1993), also argued for a continuity thesis, seeing the revolution as a product of our explanatory model, rather than of the events themselves. Even Thomas Kuhn, *The Copernican Revolution* (Cambridge, Mass.: Harvard University Press, 1957), had to acknowledge the drawn-out process of this change. R Hooykaas problematizes Copernicus' role in the scientific revolution in "The Rise of Modern Science: When and Why?," *British Journal for the History of Science* 20 (1987): 463–67.

[9]See Schuster, Chap. 3 in this volume for a similar definition. Deborah Harkness takes back the term 'science' as a legitimate one in *The Jewel House. Elizabethan London and the Scientific Revolution* (New Haven: Yale University Press, 2007).

[10]Andrew Cunningham and Perry Williams, "De-centring the 'big picture': *The Origins of Modern Science* and the modern origins of science," *British Journal for the History of Science* 26 (1993): 407–432. See Peter Dear, *Revolutionizing the Sciences: European Knowledge and its Ambitions, 1500–1700* (Princeton: Princeton University Press, 2001), for a more recent view on this question.

In other words, our explanation of the 'origins of modern science' must be more complex than Zilsel's was. While he at least had the security of believing his *explanandum* to be stable, both *explanandum* and *explanatio* must feature in our narrative. In spite of this, however, I think that the 'scientific revolution' is worth saving. While we need to take Cunningham and Williams' point seriously and be careful to avoid the presentist search for modern scientific ancestors, this does not imply that the entire enterprise is without merit. The actors themselves were aware of living in interesting times and a number of important changes took place in the investigation of nature in this period. In the 145 years between the publications of Copernicus and Newton, people interested in the Book of Nature developed new methodologies including experimentation, new attitudes towards knowledge, God, and nature, a new ideology of utility and progress, and new institutional spaces and practices.[11] They began to view the world as quantifiable, investigable, and controllable. By the end of the period, the investigation of nature was still tied to theological concerns, but also increasingly to practical ones as well, and was carried out in completely new places, for different ends, and with quite different results. Perhaps this was not the origin of modern science writ large, but it definitely had created the necessary preconditions.

However, the key to understanding this transformation must be sought in the socio-economic transformation of Europe, not simply in a metaphysical gestalt switch. Rather than seeing the development of the scientific revolution as a move "from a closed world to an infinite universe", as Koyré put it, I would argue that a sociological change was taking place in who was investigating the natural world, where these investigations took place and for what end.[12] Indeed, a crucial category of scientifically-inclined men ignored by Cunningham and Williams and downplayed by most historians of the period, the mathematical practitioners, were crucial to this transformation.[13] Mathematics was a separate area of investigation

---

[11]Shapin, *Scientific Revolution*, despite his opening caveat, does a good job of laying out some of the changes taking place that made up the scientific revolution, as more recently has John Henry, *The Scientific Revolution and the origins of modern science* (Houndmills, Basingstoke: Palgrave, 2001).

[12]Steven Shapin made a case for this new interpretation in "History of Science and its Sociological Reconstructions," *History of Science* 20 (1982): 157–211, and then, with Simon Schaffer, provided an extremely influential case study in *The Leviathan and the Air Pump: Hobbes, Boyle, and the Experimental Life* (Princeton: Princeton University Press, 1985).

[13]With some modification, I take the important classification of the more practical men in E.G.R. Taylor, *Mathematical Practitioners of Tudor and Stuart England* (Cambridge: Cambridge University Press, 1954). For modern treatment of these crucial figures, see James A. Bennett, "The Mechanic's Philosophy and the Mechanical Philosophy," *History of Science* 24 (1986): 1–28; Stephen Johnston, *Making Mathematical Practice: Gentlemen, Practitioners, and Artisans in Elizabethan England* (Ph.D. Thesis, University of Cambridge, 1994); Stephen Johnston, "Mathematical Practitioners and Instruments in Elizabethan England," *Annals of Science*, 48 (1991): 319–344; Pamela O. Long, *Artisan/Practitioners and the Rise of the New Sciences, 1400–1600* (Corvallis: Oregon State University Press, 2011); and Eric Ash, *Expertise: practical knowledge and the early modern state* (Chicago: University of Chicago Press, 2010).

from natural philosophy and those interested in mathematical issues had often tied such studies to practical applications, such as artillery, fortification, navigation, and surveying.[14] These mathematical practitioners became more important in the early modern period and provided a necessary ingredient in the transformation of nature studies to include measurement, experiment, and utility.[15] Their growing importance was a result of changing economic structures, developing technologies, and new politicized intellectual spaces such as courts, and thus relates changes in 'science' to the development of mercantilism and the nation-state. Thus, crucially, Zilsel's thesis, claiming the necessity of communication between handwork and brainwork, must now focus on these mathematical practitioners.[16] The scientific revolution was made possible by the connection established by mathematical practitioners between the more practical applications of their trade and the larger concerns of natural philosophy, often facilitated by the new political, social and cultural organization of patronage at the princely courts.

2.2 Handwork and Brainwork

The question of most importance to early twentieth-century historians of science was: why did the scientific revolution come first to Western Europe and why did it happen in the sixteenth and seventeenth centuries?[17] At least part of the answer lies in the socio-economic growth of mercantilism and the development of the arts and crafts tradition. While this is not the only reason for the changes of this period – a full answer would have to include political, cultural, and religious developments as

---

[14]Mario Biagioli, "The Social Status of Italian Mathematicians, 1450–1600," *History of Science* 27 (1989), 41–95 and *Galileo's instruments of credit: telescopes, images, secrecy* (Chicago: University of Chicago Press, 2007).

[15]James A. Bennett, "The Challenge of practical mathematics," 176–190 in *Science, Belief, and Popular Culture in Renaissance Europe*, eds. Steven Pumfrey, Paolo Rossi, and Maurice Slawinski, (Manchester: Manchester University Press, 1991). Thomas Kuhn, "Mathematical versus Experimental Traditions in the Development of Physical Science," in *The Essential Tension: Selected Studies in Scientific Tradition and Change* (Chicago: University of Chicago Press, 1977), 31–65, provides an early attempt to claim a different history for mathematics and natural philosophy.

[16]Edgar Zilsel identifies the important players as the "superior artisans". Edgar Zilsel, "The Sociological Roots of Science," *American Journal of Sociology* 47 (1942): 552–55. His superior artisans, however, are not identical to mathematical practitioners, since these artisans could not, themselves, make the move to create real scientific knowledge. They needed to work in concert with natural philosophers and it was this crucial cooperation that enabled science to emerge.

[17]H. Floris Cohen, *The Scientific Revolution. A Historiographical Inquiry* (Chicago: University of Chicago Press, 1994) sees this as one of the three main historiographical stands in this field. It is interesting to note that this was also the question that started Joseph Needham on his exploration of Chinese science. Toby Huff, *Intellectual Curiosity and the Scientific Revolution. A Global Perspective* (Cambridge: Cambridge University Press, 2010).

well – it seems fundamental. As well, it has important implications for the ultimate definition of the scientific revolution, to which I will return.

The relationship between the scholar and the craftsman, and thus between science and technology, is one that has concerned historians of science for the last 60 years. Internalists such as Rupert Hall saw at best a hierarchical relationship, with science and the scholar dictating to the craftsman and technology. At worst, this relationship was seen as incommensurable, since the two came from completely separate worlds. As Hall put it, "The scholar's function was active, to transform science; the craftsman's was passive, to provide some of the raw material with which the transformation was effected." As well, "The great discoveries of mathematical physicists were not merely over the heads of practical engineers and craftsmen; they were useless to them."[18] This was probably the majority position among historians of science from the 1950s to about 1980. On the other hand, Stillman Drake claimed that university philosophers made no contribution to the scientific revolution, but rather, men of ingenuity and practicality, like Galileo and Tartaglia, were responsible.[19] Drake himself was a man of practicality (as an investment banker) and as an autodidact, unaffiliated with university philosophers, found his hero in a like-minded individual.[20] Yet, even for Drake, Galileo was not an artisan, but rather a scientific entrepreneur. Drake was more concerned with the villains of the piece – the university scholastics who acted as intellectual gatekeepers – than with any new socio-economic explanation.

Floris Cohen's evaluation of the scholar and craftsman also reveals a hierarchical, exploitative relationship. Cohen argues, following Lynn White, that the arts and crafts tradition did influence natural philosophers like Galileo, who then turned it into something completely different.[21] White had evaluated Galileo's use of the suction pump and pendulum, two recently developed technological devices. White argued that Galileo's use of these inventions affected his choice of experiments and "makes the tonality of his new sciences historically intelligible".[22] Cohen, however, argues that Galileo's connection with craftsmen was limited to co-opting their instruments for his own more metaphysical use. The gap between rules of thumb

[18]A. Rupert Hall, "The Scholar and the Craftsman in the Scientific Revolution," in *Critical Problems in the History of Science*, ed. Marshall Clagett, (Madison: University of Wisconsin Press, 1959), 21.

[19]Stillman Drake, "Early Science and the Printed Book: The Spread of Science Beyond the Universities," *Renaissance and Reformation* 6 (1970): 43–52. Later continued in *Galileo at Work: His Scientific Biography* (Chicago: University of Chicago Press, 1978). This view of Galileo is taken up later by Matteo Valleriani in *Galileo, Engineer* (Dordrecht: Springer, 2010).

[20]Stillman Drake taught me a course on Galileo near the end of his career, and took great delight in his role as an iconoclast. He would, however, have been horrified, both intellectually and politically, to have seen any connection between his view of Galileo and Zilsel's.

[21]Floris Cohen, 346–9. See also Floris Cohen, *How Modern Science Came into the World: Four Civilizations, One 17th-Century Breakthrough* (Amsterdam: Amsterdam University Press, 2010).

[22]Lynn White, *Medieval Religion and Technology: Collected Essays* (Berkeley: University of California Press, 1978), 132.

and laws of nature was really unbridgeable and so Galileo should not be seen as a true connection between scholar and craftsman. According to Cohen, Galileo and other natural philosophers like Isaac Beeckman were ingenious in making use of materials and techniques newly available to them, but do not provide a case study to prove Zilsel's claim of a new interaction between handwork and brainwork. Having argued for the essentially exploitative nature of the early relationship, Cohen then argues that, in the late seventeenth and eighteenth centuries, scientific ideas were used to advise industrialists and produce applied scientific technology. In other words, at both moments of contact between scholar and craftsman, the scholar was clearly in a superior position. Though not as dismissive as Hall, Cohen also wants to save the scientific revolution, and science in general, for the philosopher.

As historians have sought to draw a dividing line between the scholar and the craftsmen, with their putatively different ways of knowing, they have also sought to separate pure scientific thought from sordid applied technology. Clearly this has much to do with modern issues of scientific funding, accountability, status, and hierarchy, especially in the Cold War scientific community. After all, scientists have been fighting for the right to do unfettered research for almost as long as historians have been defining the scientific revolution.[23] But does it represent any useful distinction for this early modern period? The answer must be no. Indeed, the very difficulty in discovering the difference between these two ways of knowing should provide evidence that this is the wrong question to ask. The connections between *episteme* and *techne* were often close and thus the relationship between those who knew by doing and those who knew by theorizing is extremely complex.[24] If we think of the connection between practical knowledge and theoretical knowledge as a spectrum, rather than as two discrete and incommensurable alternatives, we start to see the possibilities of interaction between the two. While hands-on estimates at one end contrasts sharply with laws of nature at the other, the gradations in between can allow individuals and groups of individuals to interact and to use different modes of thought at different times. For example, mathematicians such as Henry Briggs, well-versed in the more transcendental theories of their discipline, could choose to ignore these for the real-life applicability of a theoretically-suspect

[23]There is much modern literature on the importance of pure research, e.g. Henry Etzkowitz, Andrew Webster, and Peter Healey, eds., *Capitalizing Knowledge: New Intersections of Industry and Academia* (New York: University of New York Press, 1998) and Linus Pauling, "Chemistry and the World Today. An invitation – and a warning – to private industry to come to the aid of basic research," *Engineering and Science Monthly* XIII (1), October 1949: 5–8. J.J. Thomson articulated this much earlier when he said, "Research in applied science leads to reforms, research in pure science leads to revolutions." Quoted in J.D. Bernal, *Science in History* (London: Watts & Co., 1954), 42. Or see Vannevar Bush, *Science, the Endless Frontier* (Washington, D.C.: US. G.P.O., 1945) who calls for practically motivated research.

[24]Bennett, for example, points out the important cross-over between practical mathematicians and natural philosophers in the seventeenth century, although he provides no mechanism for this cross-over. Bennett, "Challenge of practical mathematics".

calculus.[25] Similarly, William Gilbert could use practical studies of compass dip to make larger philosophical arguments about the composition of the earth.[26] It is within this slippage from one way of knowing to another that we find some of the clues to the development of a 'new science' in the sixteenth and seventeenth centuries.

This relationship between scholar and craftsman was first articulated in early Marxist interpretations of the scientific revolution. Both Boris Hessen and Edgar Zilsel claimed a connection between the growing technologies and economic innovations of early modern Europe and the development of new scientific models. We need to reexamine the Hessen thesis and particularly the Zilsel thesis in order to understand this extremely important connection between theory and practice.

2.3 Hessen and Zilsel

The Hessen thesis, a rather naive application of Marx's historiography to the history of the scientific revolution, is definitely the more notorious of the two. Boris Hessen, a Soviet physicist, presented his thesis at the International Congress of the History of Science and Technology in London in 1931.[27] Hessen was prominent in Soviet circles until his disappearance in 1934; he is thought to have died in one of the Stalinist purges of the 1930s. Loren Graham argues that Hessen's paper, "The Social and Economic Roots of Newton's 'Principia'", was an attempt to separate the value of Newton's work from its theological and anti-material roots and therefore by extension, to make the same claim for Einstein.[28] Hessen was a supporter of Einstein's theory of relatively, a suspect position in Soviet circles. Given Hessen's fate, this ploy was clearly unsuccessful. For our purposes, however, it resulted in an interesting articulation of the relationship between materialism and the scientific revolution. In his article, Hessen argues that Newton developed his theories because of the newly bourgeois society of England, and because of the mechanical engines being created by craftsmen.[29] Hessen itemized each

[25] Katherine Neal, *From discrete to continuous: the broadening of number concepts in early modern England* (Dordrecht: Kluwer Academic Publishers, 2002).

[26] Stephen Pumfrey, *Latitude and the Magnetic Earth* (Cambridge: Icon Books, 2003).

[27] Pamela Long, *Artisan/Practitioners* (2011), discusses Hessen as the first major Marxist analyst of the scientific revolution. For information about Hessen's life, see P.G. Werskey, "Introduction," in Nikolaĭ Bukharin, *Science at the Cross Roads* (London: Kniga Ltd., 1931) xv–xvi, xx–xxi and Loren Graham, "Socio-political roots of Boris Hessen: Soviet Marxism and the History of Science," *Social Studies of Science* (1985), 705–722. Graham puts Hessen into a larger context in *Science in Russia and the Soviet Union: a short history* (Cambridge: Cambridge University Press, 1993).

[28] Graham,"Socio-political roots," 706.

[29] Boris Hessen, "The Social and Economic Roots of Newton's 'Principia'," *Science at the Cross Roads* (London, 1931).

technological need of the seventeenth century and claimed that the particular scientific investigations of the period developed as a direct result of technological need. For example, seventeenth-century English merchants needed to increase the tonnage of their ships and therefore needed to understand hydrostatics, which was one of the areas that scientists of the day were investigating. Likewise, mechanical devices so important to early capitalist technology encouraged and needed scientific explanations of mechanical laws. This is why, according to Hessen, scientists of the seventeenth century did not produce thermodynamics – without the steam engine as inspiration and a site of observation and experimentation, scientists could not develop such a theory. Science for Hessen was completely dependent, first on the economic structure, and second on the technological level of the society.

Hessen's paper followed a strict Marxist analysis, with relatively little historiographical sophistication, so it is no surprise to find that the reaction to his work was almost completely negative.[30] This was, of course, aided by deteriorating relations between the Soviet Union and the west, as well as growing fears of Communism and Marxism in general. Indeed, it is ironic that Hessen's work gained him as few supporters in the Soviet Union as in the West and were it not for the publication of his article in a book printed in England and the resonance of his ideas with a few other western scholars, he would have dwindled into complete obscurity. One such scholar was Edgar Zilsel and although his work initially suffered a fate similar to Hessen's, his relationship to the Vienna Circle and the exiled Frankfurt School ensured that he would be considered by western scholars, even if long after his death.

Edgar Zilsel was an Austrian Jewish Marxist, a marginal member of the Vienna Circle and of the New York version of the Institute of Social Research (the Frankfurt School in exile).[31] He was a positivist who believed that knowledge, life, and education should be unified, exemplified by his involvement with the *volkshochschule* movement (adult education). Part of his marginalisation with respect to the Vienna Circle was due to his commitment to practice.[32] It is interesting that his historical theory of scientific development emphasized that same interaction between theory and practice.

Zilsel devoted much of his academic research in the latter part of his life to the question of why the scientific revolution took place when and where it did. For Zilsel this was a quest for the beginning of true science and much of his energy was devoted to identifying the sociological barriers to its emergence, finally overcome in the late sixteenth and seventeenth centuries. In "The Sociological Roots of Science" (1942), he claimed that the emergence of modern science was a sociological process – with the transition from feudalism to early capitalism

[30] F. Cohen, 331–3.

[31] Deiderick Raven, Wolfgang Krohn and Robert S. Cohen, eds., *Edgar Zilsel: The Social Origins of Modern Science*, (Dordrecht, Boston and London, 2000), xxxix–xlvii.

[32] *Ibid.*, xxxiv–vi.

and the growth of technology in the towns as crucial to this process.[33] But rather than seeing the new science developing from the needs of technology, as Hessen had done, or even because of the ability to observe the new machines, Zilsel believed that the crucial transformation took place because of the coming together of scholars and craftsmen. Zilsel argued that before the emergence of modern science, three different groups of intellectual workers kept separate the necessary ingredients for this transformation. Scholastics, basing their work on reason and tradition rather than on the formation of natural laws, were the university-based intelligentia. Humanists were most concerned with language and antiquity. Only the final group, the superior artisans, provided the experience of nature that could produce experimental method and a belief in progress necessary to modern science. As long as these three were kept separate, no transformation in natural knowledge could occur, according to Zilsel. It was only with the development of towns and cities, due to the emergence of capitalism, that these groups could come together to create the 'new science'.

This rapprochement was possible for several reasons. First, this new group of superior artisans (artist-engineers, instrument makers, surveyors, and navigators) began to emerge, separate from both university scholars and humanist literati, as well as from the older guild structures. This new group developed an experimental method which placed value on empiricism, quantification, and cooperation, while still participating in the growing capitalist and nationalist enterprise. As Zilsel put it:

> When the seamen of the sixteenth century went to sea, they laid the foundation-stone of the British Empire and when they returned and made compasses, of modern experimental science.[34]

Second, the status of these practitioners began to rise, in part due to changing economic circumstances. Finally, beginning in about 1550, this group began to come together with the more theoretical natural philosophers to produce an entirely new way of interpreting the natural world.[35] This communication was possible because of the new individualism of early capitalist enterprises and the freer atmosphere of the early modern towns.[36] Thus, for Zilsel, the crucial century was the sixteenth where for Hessen it had been the seventeenth. As well, although the development of capitalism and its related technologies is extremely important to Zilsel's argument, his explanation for change is socio-cultural as well as economic. Zilsel was more concerned with these superior artisans (whom he saw both bringing together theory and practice themselves and especially providing new information and method for

[33]Edgar Zilsel, "Sociological Roots of Science." See also "Copernicus and Mechanics," *Journal of the History of Ideas* 1 (1940): 113–118, reprinted in *Roots of Scientific Thought*, ed., Philip P. Wiener and Aaron Noland (New York: Basic Books, 1957), 276–280.

[34]Edgar Zilsel, "The Origins of Gilbert's Scientific Method," *Journal of the History of Ideas* 2 (1941), reprinted in *Roots of Scientific Thought*, 241.

[35]Zilsel, "Sociological Roots," 554. In "Gilbert" he dates this rapprochement to about 1600.

[36]Zilsel, "The Genesis of the Concept of Scientific Progress," *Journal of the History of Ideas* 6 (1945): 325–49, reprinted in *Roots of Scientific Thought*, 257–8.

natural philosophers such as William Gilbert) and the values associated with this cultural moment, than with production processes as Hessen had been. He saw the importance of cooperation and communication, as well as the development of the idea of progress. Zilsel's scientific revolution was based primarily on an economic revolution, but also included social, cultural, political, and discursive elements.

There are a number of problems with this thesis. While Zilsel posits capitalism and the growth of towns as his mechanism for change, this is clearly not sufficient. The towns had been developing for several hundred years, during which time guild structures had been changing as well. As Pam Long has shown, late medieval trade guilds were not as secretive nor as hidebound as Zilsel and other earlier commentators once thought, and innovation and individualism were thus in place much earlier.[37] Mercantilism is certainly a new development in the fifteenth and sixteenth centuries, but is not synonymous with Zilsel's capitalism. Although economics is very important, it cannot alone explain the scientific revolution, since many of the social and cultural changes necessary to the growth in power and status of the 'superior artisans' owed as much to the military, political, and religious developments of the elite as to trade with distant lands. Zilsel also fails to explain why or how this new information, method, and ideology were important to the world in which they were introduced. Zilsel simply asserts that experiment, an epistemological belief in natural laws, and a belief in progress mattered, rather than supplying any sociologically sophisticated explanation for why natural philosophers ultimately took these methods and ideologies on board.

At his untimely death, Zilsel's work remained unfinished and, although he had published in English and in important American venues, his thesis was soon ignored.[38] A full evaluation of the causes of this eclipse is beyond the scope of this chapter, but had much to do with Cold War suspicion of Marxist or even materialist explanations of scientific breakthroughs. Historians of science in the years after World War II were concerned that science remains untainted by outside influences, in order to be available to serve mankind. Zilsel's ideas were probably dangerous to contemplate, which may account for the indirect nature of Rupert Hall's virulent attack, as well as the virtual abandonment of Zilsel by the Anglo-American history community thereafter.[39]

---

[37] Pamela Long, *Openness, secrecy, authorship: technical arts and the culture of knowledge from antiquity to the Renaissance* (Baltimore: Johns Hopkins University Press, 2001). See also William Eamon, *Science and the Secrets of Nature: Books of Secrets in Medieval and Early Modern Culture* (Princeton: Princeton University Press, 1994).

[38] Raven and Krohn, "Introduction."

[39] Hall's "Scholar and Craftsman," although never mentioning Zilsel in text or footnote, was clearly an attack on Zilsel's thesis. Robert Merton, "Science, Technology, and Society in Seventeenth-Century England," *Osiris* 4:2 (1938): 360–632, was influenced by both Hessen and Zilsel, but the second half of this long essay which spoke about the importance of capitalism for the development of modern science was largely ignored in favour of the Weber-influenced religious argument of the first section. Paolo Rossi, especially with *Philosophy, Technology and the Arts in Early Modern Europe* (London: Harper and Row, 1970), shows that a pro-Zilsel camp continued to exist in Europe.

2.4 Utility and Ben-David's Scientistic Society

A few scholars did continue to evaluate changing cultural and sociological attitudes towards science in this early modern period, with an eye to an explanation of the scientific revolution. Perhaps the two most important ideological strands could be called 'Baconian ideology', that is a belief in the utility of natural knowledge, and the growth in what Ben-David called a 'scientistic society'.[40] The first interpretation claims that Francis Bacon articulated and then inspired a belief in the power and utility of knowledge about nature, leading to a belief in progress and a push to such cooperative and practical scientific ventures as the founding of the Royal Society.[41] Part of the problem with such an interpretation is that these so-called 'Baconian' ideals of progress, power and utility were in common circulation in the sixteenth century, merely extolled rather than invented by Bacon. As Zilsel himself said, "Manifestly, the idea of science we usually regard as 'Baconian' is rooted in the requirements of early capitalistic economy and technology".[42] Thus, Bacon cannot somehow stand as a mechanism for change. Likewise, the Royal Society owes less to Bacon as a founding father than to changing social, cultural and economic realities in London elite society.[43] Equally, Ben-David's suggestion that Western European society became more and more interested in and sympathetic towards a scientific model and its knowledge is really an argument for legitimation rather than explanation. Ben-David offers an invaluable evaluation of how science became an increasingly important factor in European intellectual life in the eighteenth and nineteenth centuries, but tells us very little about how it got that way in the sixteenth and seventeenth centuries. Shapin, for example, suggests that this scientistic society was a very long time in coming. In the sixteenth and

---

[40]For Bacon, see Julie Robin Solomon and Catherine Gimelli Martin, eds., *Francis Bacon and the Refiguring of Early Modern Thought: Essays to Commemorate the Advancement of Learning (1605–2005)* (Aldershote: Ashgate, 2005), especially Jerry Weinberger, "Francis Bacon and the Unity of Knowledge: Reason and Revelation," Also, R. Julian Martin, *Francis Bacon, The State, and the Reform of Natural Philosophy* (Cambridge: Cambridge University Press, 1992). Joseph Ben-David, *The Scientist's Role in Society: A Comparative Study* (Englewood Cliffs, N.J.: Prentice Hall, 1971). On Ben-David, see F. Cohen, 367–73. It is interesting that F. Cohen feels Ben-David's is the most plausible 'external' explanation for the scientific revolution, since it never delves into the scientific ideas at all.

[41]This interpretation began with Enlightenment thinkers such as d'Alembert (F. Cohen, 22–23), and includes more modern historians such as R.F. Jones, *Ancients and Moderns: A Study of the Rise of the Scientific Movement in Seventeenth-Century England* (Berkeley: University of California Press, 1936) and Charles Webster, *The Great Instauration* (New York, 1976; Oxford: Oxford University Press, 2002).

[42]Zilsel, "Genesis," 272. However, Zilsel believed in the importance of Bacon's influence: "The concept of scientific progress was known before him, the ideal of the progress of civilization begins only with Bacon." *Ibid.*

[43]Steven Shapin, "The House of Experiment in Seventeenth-Century England," *Isis* 79 (1988): 373–404, and *A Social History of Truth: Civility and Science in Seventeenth-Century England* (Chicago: University of Chicago Press, 1994).

seventeenth centuries natural philosophers did not command instant respect, nor did they choose scientific exactness over continued civility and gentle conversation.[44]

Despite these caveats, however, a greater and greater stress on practicality and utility, combined with a greater knowledge of and interest in science and nature on the part of a larger segment of society, were important elements to the changing attitude towards nature in the early modern period. Both were influenced by the development of mercantilism, the growth of towns, and changing governing needs and strategies. Indeed, the claim of utility and power over nature, expressed by mathematical practitioners as they aspired to court positions, helped create a sense of the importance of studying the natural world and so the two interpretations are related, especially at the level of princely attitudes and concerns. The scientific revolution combined new attitudes to nature and power, new players searching for status, and new possibilities opened up by economic, political, and cultural change in the early modern period.

2.5 The 'Scientific Revolution' and Mathematical Practitioners

The sixteenth century was a time of dislocation for natural philosophers. As the Roman Catholic Church lost its professed monopoly on Truth, so too did university scholastics.[45] Although universities continued to be important, their clientele began to change.[46] At the same time, a window of opportunity was created, especially through patronage in the princely courts. Because of this, different things began to be valued. Rather than syllogistic logic and theological subtleties, princes wanted spectacle, power, and wealth. Therefore, natural philosophers who were practical (or claimed to be) were valued.[47]

Thus, the line between court natural philosopher and (mathematical) practitioner was not clear. Authors of practical treatises dedicated or presented their works to patrons and princes in order to raise the status of the practitioner and at the

[44]Steven Shapin, "'A Scholar and a Gentleman': The Problematic Identity of the Scientific Practitioner in Early Modern England," *History of Science* 29 (1991), 279–327, and *Social History of Truth*, ch. 5.

[45]For an interesting assessment of this relationship, see Andrew Weeks, *Paracelsus. Speculative Theory and the Crisis of the Early Reformation* (Albany: State University of New York Press, 1997).

[46]James K. McConica, ed., *The Collegiate University. The History of the University of Oxford*. Vol. 3 (Oxford: Oxford University Press, 1986), 1–68.

[47]For studies of patronage of science, see Bruce T. Moran, ed., *Patronage and Institutions. Science, Technology, and Medicine at the European Court* (Rochester, N.Y.: Boydell Press, 1991); Paula Findlen, *Possessing Nature: Museums, Collecting, and Scientific Culture in Early Modern Italy* (Berkeley: University of California Press, 1994); Pamela H. Smith, *The Business of Alchemy: Science and Culture in the Holy Roman Empire* (Princeton: Princeton University Press, 1994).

same time legitimize empirical knowledge. For example, alchemists, astrologers, and builders of curious devices were valued both for the status they brought to a court and for the practical results they might achieve. Portuguese pilots, who in the early fifteenth century were beneath the notice of king or court, were made Royal Cosmographers in the sixteenth century.[48] Most natural philosophers attached to princely courts gained their reputations both for intellectual acuity and for practical applications. For example, Johannes Kepler and John Dee cast horoscopes for Rudolf and Elizabeth respectively. Dee advised Elizabeth on the most propitious day for her coronation, as well as consulting with navigators searching for a north-west passage.[49] Likewise, Galileo's activities as a courtier were both esoteric and applied.[50] These men walked a fine line between theory and practice, since all three were interested in large philosophical systems and desired court patronage not simply for creating improved telescopes or new armillary spheres. But as Dee's case makes clear, monarchs wished results more tangible than angelic conversations and all investigators of the natural world with court connections were compelled on occasion to dance for their supper. Even investigators less directly attached to courts, such as geographers William Gilbert or Richard Hakluyt, combined an interest in more theoretical cosmographical issues with direct practical results. Hakluyt wished to construct a complete image of the globe, but presented his work as an imperial and intensely practical project.[51] Gilbert spoke of mining and navigation, while constructing a new theory of earthly magnetism.[52]

Thus, utility of knowledge was of prime importance in this new regime, as was direct connection between scholar and craftsman (or at least scholarly and craft ideas). This was more complicated, however, than just the development of a proto-capitalist economy. Mercantilism grew and flourished in this period, and its expansion was fundamental to much of the new political organization in Europe. Nevertheless, utility was a rhetorical position for many, rather than an expression of direct application of new philosophical ideas to the marketplace. For one thing, a new courtly culture was developing, and for many entrée to this culture was

[48]Jerry Brotton, *Trading Territories. Mapping the Early Modern World* (London: Reaktion Books, 1997), 51–65.

[49]Martin Frobisher, Richard Chancellor, Pet, Jackman, Humphrey Gilbert and Sir Walter Ralegh all took Dee's advice about navigation and strategy. John Dee, *The Private Diary of Dr. John Dee*, ed., J.O. Halliwell (London, 1842), esp. 18, 33.

[50]Mario Biagioli, *Galileo Courtier: The Practice of Science in the Culture of Absolutism* (Chicago: University of Chicago Press, 1993).

[51]Richard Hakluyt, *The Principal Navigations, Voiages, Traffiques and Discoveries of the English Nation*, 3 Vols. (London, 1598–1600). This contrasts with *Divers Voyages touching the discoverie of America, and the Ilands adiacent unto the same, made first of all by our Englishmen* (London,1582). Zilsel names Hakluyt as one of those crossing the divide between scholar and craftsman in "Gilbert," 246.

[52]Pumfrey, *Latitude*, and Bennett, "Practical mathematics."

through a claim to utility. [53] Likewise, the entrepreneurial mathematicians springing up across Europe were at least as concerned with appearing to be useful as they were with the real applications of their ideas and expertise. In other words, Zilsel's straight-forward Marxist interpretation does not apply here without amendment. People needed to earn a living, but money alone was less important to many of these new natural investigators than cultural capital. Utility was a discourse and an ideology, and the economic possibilities were more related to the sale of expertise than to any get rich quick schemes.[54]

In the seventeenth century, the relationship between natural philosophers, craftsmen, and patrons began to change. As the status of the natural philosopher began to improve, he no longer needed to justify himself in terms of application of his knowledge. As well, he began to see the importance of separating himself from the rude mechanicals and so even mathematical practitioners began to claim a more elite audience for their work. This separation of the elite from the populous occurred not just in natural philosophy, but in many cultural venues in the seventeenth century. This was the period when the upper classes were withdrawing more generally from popular culture, for example from carnival, Christmas celebrations, and other revelries.[55] As is illustrated in the final scenes of *A Midsummer-Night's Dream* and *Love's Labour's Lost*, the elite were more content to watch common festivities than to take part, allowing Bottom and his 'rude mechanicals' to act out Pyramus and Thisbe in the first, and the workers to enact the Nine Worthies in the latter.[56]

In a natural philosophical setting, we can see this separation most clearly in Francis Bacon's *Solomon's House*.[57] While some have interpreted this ideal laboratory as a true democratization of scientific knowledge, Bacon's aim here is rather to control the production of truth. As Julian Martin has shown us, Bacon distrusted the disorder of the common people and, far from democratizing science with his plan for the participation of all, relegated the mechanicals to a lower and lower position – as hewers of wood and drawers of water. The fact-gathers for Solomon's House were not to understand the overall process or theory, but gather their individual bits of information, leaving only the approved philosophers, men

[53] Katherine Neal, "The Rhetoric of Utility: Avoiding Occult Associations for Mathematics through Profitability and Pleasure," *History of Science 37 (2)* (1999): 151–78, talks about some attempts to make mathematics seem useful.

[54] This is also true for that interesting seventeenth-century innovator, the 'projector'. See Charles Webster, *The Great Instauration: Science, Medicine and Reform 1626–1660* (2nd ed. with new Preface, Oxford, 2002), for a discussion of the projectors around the Hartlib Circle.

[55] Peter Burke, *Popular Culture in Early Modern Europe* (London: Temple Smith, 1978).

[56] William Shakespeare, *A Midsummer-Night's Dream*, Act 5, scene 1; *Love's Labour's Lost*, Act 5, scene 2.

[57] Francis Bacon, *The New Atlantis* (London, 1627). An easily accessible version is Andrew Ede and Lesley Cormack, eds., *A History of Science in Society: A Reader* (Peterborough: Westview Press, 2007), 157–161.

like Bacon himself, to draw conclusions and create axioms.[58] Bacon, a true elitist, wanted to ensure that only those as trustworthy as himself would have any say in theory. Thus, practitioners and craftsmen generally had lost status, even as natural philosophers purported to be practical.

By the time of the establishment of the Royal Society and the Académie des Sciences, those studying the natural world had ceased to communicate intimately with more practical men or ideas. Those who had successfully created the 'new science' had constructed social barriers to divide theory from practice and thus had begun the sense of separation that would be important to modern scientific ideas of pure research. The scientific revolution was made possible by communication along the spectrum of handwork to brainwork, but was completed by socially severing that connection.

2.6 The Case of English Geography

Nowhere was this blend of utility, curiosity about nature, and political agendas more overt than in the study of geography. Geographers were encouraged at many princely courts in early modern Europe, both because such princes had an economic and intellectual interest in the newly discovered world and because such mathematical practitioners could add lustre and importance to their courts. At the same time, merchants had an economic and political interest in understanding the globe, as well as a growing desire to seem geographically astute. The relationship between intellectual geographical interests and political and economic concerns was developing all over Europe. Thus geography provides a particularly good example of the transformation of science taking place in this period. The study of geography, usually by mathematical practitioners, necessitated an interaction between theory and practice. It was encouraged by new mercantile concerns and equally by new political and cultural realities. Thus, the missing mechanisms of Zilsel's explanation are present in this case study.

In sixteenth-century England, geography was a flourishing area of investigation. It was studied as part of the arts curriculum at both Oxford and Cambridge and therefore made up part of the worldview of most educated gentlemen and merchants.[59] The study of geography included a mathematical model of the earth, descriptions of its distant lands and inhabitants, and the local history of more immediate surroundings, what I have elsewhere labelled mathematical geography, descriptive

[58]Martin, *Francis Bacon*, 136–39. Shapin and Schaffer, *Leviathan*, take this in a slightly different direction by arguing for the importance of witnesses of a certain status in order to create 'matters of fact'.

[59]See Lesley B. Cormack, *Charting an Empire. Geography at the English Universities 1580–1620* (Chicago: University of Chicago Press, 1997), 17–47, for a full treatment of the place of geography in the university curriculum.

geography, and chorography.[60] Because it relied on geographers of antiquity, such as Ptolemy and Strabo, to provide a backbone for modern investigation, geography was a discipline that used the methods of the humanists and the tradition of university scholars. Equally, geography was a study inspired by and reliant on new discoveries, voyages, and travels and so was integrally connected to the testimony and experience of practical men. Thus, geography existed as a point of contact for theoretical university scholars and practical men of affairs, and so provides a wonderful Zilselian moment in which to appraise the changing scientific world.

Geography embodied that dynamic tension between the world of the scholar, since geography was an academic subject legitimated by its classical, theoretical, and mathematical roots, and the world of the artisan, since it was inexorably linked with economic, nationalistic, and practical endeavours. It provided a synthesis that enabled its practitioners to move beyond the confines of natural philosophy to embrace a new ideal of science as a powerful tool for understanding and controlling nature. The usefulness of geographical study was of paramount importance to the new men attending the universities in ever greater numbers and it was this concept of utility to the state and to the individual that drove these new university men to investigate and appreciate geography.[61] The geographical community, then, was a wide-ranging group, with many different concerns and goals, but with a desire to be useful to the nation and to their own self-interest and a vision of England as an increasingly illustrious player on the world stage.

The English geographical community was complex, due in large part to its necessarily close connection between handwork and brainwork. Even the most theoretical geographer required the information and insight of navigators, instrument makers, cartographers, and surveyors in order to understand the terraequeous globe.[62] This can be seen in the work of Richard Hakluyt, who used sailors' tales to construct a description of the world and England's role in its discovery, and who in *Principal Navigations* created a predominantly practical document with important theoretical insights. Edward Wright, a serious mathematical geographer whose first-hand experience on voyages of discovery deeply affected his research program, also provides an important example of someone who mediated between theory and practice, as we will see. Equally, the collaboration between John Dee, a university-trained mathematician and geographer, and Henry Billingsley, a London merchant, in the 1570 translation of Euclid indicates the fruitful exchange between the life of the mind and that of the marketplace.[63] Dee's career provides a particularly

---

[60] See Lesley B. Cormack, "'Good Fences Make Good Neighbors': Geography as Self-Definition in Early Modern England," *Isis* 82 (1991): 639–661, for a description of the different types of geography studied in sixteenth- and seventeenth-century England.

[61] See Lawrence Stone, "The Educational Revolution in England, 1560–1640," *Past and Present* 28 (1964): 41–80, for an evaluation of the growing numbers of new men at the universities in this period.

[62] See Chap. 4 in this volume for a fuller discussion of the role of instrument-makers.

[63] John Dee, *The Mathematical Preface to the Elements of Geometrie of Euclid of Megara* (London, 1570). It is important to note that neither was attached to a university.

telling example of the importance of the theoretical-practical spectrum. Dee would probably have identified himself as a natural philosopher and certainly worked throughout his life to create a new theoretical worldview as well as to achieve a higher social status. Yet he was engaged much of the time in more practical mathematical pursuits, especially astronomical and geographical ones.[64] He advised most navigators setting out on north-west or north-east voyages, devised map projections and navigational instruments and wrote position papers for the Privy Council on the political ramifications of English geographical emplacement.[65] Thus Dee, like many other mathematical practitioners, developed multiple and overlapping roles as scholar, craftsman, and statesman.

As was the case with Dee, many English geographers combined a university education with court and mercantile experience. A clear majority of English geographers in the late sixteenth and early seventeenth centuries attended one of the two universities and thus Zilsel's dismissal of the universities cannot be accurate.[66] A surprising number were also participants at either the Elizabethan or Jacobean courts. In order to be welcomed, they had to combine technical knowledge, political savvy and transcendental knowledge. No sea captains need apply, but equally, being a mere university don was not appropriate either. In other words, the intersection that Zilsel sees taking place between the craftsman and the scholar took place within the person of the court geographer.[67] Thus, patronage, at the court (with its political, cultural, and economic implications) was often the place – and reason – for the coming together of theory and practice.[68]

There are many examples of this interaction. Zilsel was right to draw our attention to the connections between William Gilbert and Robert Norman, for

---

[64]For the natural philosophical work, see Nicholas Clulee, *John Dee's Natural Philosophy: Between Science and Religion* (London: Routledge, 1988). For his practical advising, see William H. Sherman, *John Dee: The Politics of Reading and Writing in the English Renaissance* (Amherst: University of Massachusetts Press, 1995).

[65]In many ways, Dee is the English equivalent of Galileo, providing a cross-over from mathematical practitioner to court natural philosopher. It is no surprise, however, that Zilsel did not mention him, since his magical heritage, made famous by Frances Yates, *The Rosecrucian Enlightenment* (London: Routledge and Kegan Paul, 1972), among others, discounted him in Zilsel's mind as a true scientist. See Deborah E. Harkness, *John Dee's conversations with angels: Cabala, Alchemy and the end of nature* (Cambridge: Cambridge University Press, 1999). For new takes on Dee, see (inter alia) Stephen Clucas, *John Dee: interdisciplinary studies in English Renaissance thought* (Dordrecht: Springer, 2006).

[66]See Cormack, *Charting an Empire*, for the university and further careers of English geographers in this period. Zilsel, "Sociological Roots," 548, for a dismissal of the universities.

[67]Cormack, "Twisting the Lion's Tail: Practice and Theory at the Court of Henry Prince of Wales," in *Patronage and Institutions*, ed., Bruce Moran (Rochester, NY: Boydell Press, 1991) 67–84, discusses the presence of geographers at the court of Henry, Prince of Wales, and the resulting research program in imperial geography.

[68]Joint stock companies were another locale for this exchange and need to be examined in depth for this contribution. See Richard Hadden, *On the Shoulders of Merchants* (Albany: State University of New York Press, 1994); Richard Helgerson, *Forms of Nationhood. The Elizabethan Writing of England* (Chicago: University of Chicago Press, 1992), 151–181, on Hakluyt and the merchants.

instance. Two geographers who combined the life of the scholar with that of the practitioner were Edward Wright and Thomas Harriot.[69] Both were university-educated men, who there learned the classical foundations of their subject, as well as recent discoveries and theories. But these two were not isolated or traditional scholastics. Both went on prolonged voyages of discovery and learned navigation and its problems from the rude mechanicals and skilled navigators they encountered. They went beyond this practical knowledge, however, to try and formalize the structure of the globe and the understanding of the new world. Both were connected with important courts and patrons, and both used the cry of utility and imperialism to argue the need for geographic knowledge.

Edward Wright, the most famous English geographer of the period, was educated at Gonville and Caius College, Cambridge, receiving his B.A. in 1581 and his M.A. in 1584. He remained as a fellow at Cambridge until the end of the century, with a brief sojourn to the Azores with the Earl of Cumberland in 1589.[70]

In 1599 Edward Wright translated Simon Stevin's *The Haven-finding Arte* from the Dutch.[71] In this work Stevin claimed that magnetic variation could be used as an aid to navigation in lieu of the calculation of longitude.[72] He set down tables of variation, means of finding harbours with known variations, and methods of determining variations. In his translation Wright called for systematic observations of compass variation to be conducted on a world-wide scale,

> that at length we may come to the certaintie that they which take charge of ships may know in their navigations to what latitude and to what variation (which shal serve in stead of the longitude not yet found) they ought to bring themselves.[73]

Wright's work demonstrates a close connection between navigation and the promotion of a 'proto-Baconian' tabulation of facts meant both for practical appli-

[69] A. J. Apt, "Wright, Edward (*bap.* 1561, *d.* 1615)," and J. J. Roche, "Harriot, Thomas (*c.*1560–1621)," in *Oxford Dictionary of National Biography*, ed., H. C. G. Matthew and Brian Harrison (Oxford, 2004); online edition, ed. Lawrence Goldman, October 2006, www.oxforddnb.com/view/article/30029 and 12,379 (accessed December 7, 2007).

[70] As a result of this voyage, Wright wrote *The Voyage of the Right Honorable the Earle of Cumberland to the Azores*, which was later printed in 1599 and then reprinted by Richard Hakluyt, "written by the excellent Mathematician and Enginier master Edward Wright," *Principal Navigations, Voiages, Traffiques and Discoveries of the English Nation* (London, 1598–1600), II.2: 155 [misnumbered as 143]-168. M. B. Hall, *The Scientific Renaissance 1450–1630* (London: Harper and Brothers, 1962), 204; David W. Waters, *The Art of Navigation in England in Elizabethan and Early Stuart Times* (New Haven: Yale University Press, 1958), 220; J.W. Shirley, "Science and Navigation in Renaissance England," in *Science and the Arts in the Renaissance*, eds., John W. Shirley and F. David Hoeniger (Washington, D.C.: Folger Shakespeare Library, 1985), 81; all cite this trip to the Azores as the turning point in Wright's career, since it convinced him in graphic terms of the need to revise completely the whole navigational theory and procedure.

[71] Edward Wright, *Haven finding art* (London, 1599), and E.G.R. Taylor, *Mathematical Practitioners,* #100.

[72] Edward Wright, *Haven finding art*, 3. Bennett marks the relationship between magnetism and longitude as one of the important sites of the scientific revolution. "Practical mathematics," 186.

[73] *Ibid.*, "Preface," B3a, and Waters, *Art of Navigation*, 237.

cation and scientific advancement. Here appears the foundation of an experimental science, grounded in both practical application and theoretical mathematics, quite separate from any more traditional Aristotelian natural philosophy or Neoplatonic mathematics. Unfortunately, Wright's scheme was not entirely successful. By 1610, in his second edition of *Certaine Errors of Navigation*, Wright had constructed a detailed chart of compass variation – but he had also become more hesitant in his claims concerning the use of variation to determine longitude.[74]

Wright‘s greatest achievement was *Certaine Errors in Navigation* (1599), his appraisal of the problems of modern navigation and the need for a mathematical solution. In this book, Wright explained Mercator‘s map projection for the first time, providing an elegant Euclidean proof of the geometry involved. He also published a table of meridian parts for each degree, which enabled cartographers to construct accurate projections of the meridian network, and offered straightforward instructions on map construction.[75] As well, he constructed his own map using this method. Wright's work was the first truly mathematical rendering of Mercator's projection and placed English mathematicians, for a time, in the vanguard of European mathematical geography. It was equally significant for the close communication it claimed and required of theoretical mathematicians and practical navigators.

At about the turn of the century, Wright moved from Cambridge to London, where he established himself as a teacher of mathematics and geography. At about the same time, he contributed to Gilbert‘s work on magnetism, providing a practical perspective to Gilbert's more natural philosophical outlook.[76] He created a world map using Mercator's techniques and probably aided in the construction of the Molyneux globes.[77] In the early seventeenth century, he is said to have become a tutor to Henry, Prince of Wales, (elder son of James) a claim strengthened by Wright's dedication of his second edition of *Certaine Errors* to Henry in 1610.[78] Upon becoming tutor, Wright

> caused a large sphere to be made for his Highness, by the help of some German workmen; which sphere by means of spring-work not only represented the motion of the whole celestial sphere, but shewed likewise the particular systems of the Sun and Moon, and their

---

[74]Edward Wright, *Certaine Errors in Navigation* 2nd edition (1610), sigs. 2P1a-8a, and Waters, *Art of Navigation*, 316.

[75]Wright, *Certaine Errors in Navigation* (London, 1599), sigs. D3a-E4a, and E.G.R. Taylor *Mathematical Practitoners*, #99.

[76]Pumfrey, *Latitude*, 175–181.

[77]Helen M. Wallis, "The Molyneux Globes," *B.M. Quarterly* (1952): 89–90; and "'Opera Mundi': Emery Molyneux, Jodocus Hondius and the first English Globes," *Theatrum Orbis Librorum*, eds. Ton Croiset van Uchelen, Koert van der Horst and Günter Schilder (Utrecht: HES Publishers, 1989), 94–104. Lesley B. Cormack, "Glob(al) Visions: Globes and their Publics in early modern Europe," 138–156 in *Making Publics in Early Modern Europe: People, Things, Forms of Knowledge*. Eds. Paul Yachnin and Bronwen Wilson (New York: Routledge University Press, 2009).

[78]Edward Wright, *Certaine Errors in Navigation* 2nd edition (1610), sigs. *3a-8b, X1–4; *Dictionary of National Biography*, Vol. 21, 1016; Thomas Birch, *Life of Henry, Prince of Wales, Eldest Son of King James I* (London, 1760), 389.

> circular motions, together with their places, and possibilities of eclipsing each other. In it was a work by wheel and pinion, for a motion of 171000 years, if the sphere could be kept to long in motion.[79]

Henry had a decided interest in such devices and rewarded those who could create them.[80] As well, Wright designed and constructed a number of navigational instruments for the Prince and prepared a plan to bring water down from Uxbridge for the use of the royal household.[81] In or around 1612, Wright was appointed librarian to Prince Henry, but Henry died before Wright could take up the post.[82] In 1614, Wright was appointed by Sir Thomas Smith, governor of the East India Company, to lecture to the Company on mathematics and navigation, being paid £50 per annum by the Company.[83] There is some speculation as to whether or not Wright actually gave these lectures, since he died in the following year.

Wright thus provides a nice example of a mathematical practitioner who provided both intellectual and social connections between theory and practice. He was university-trained and worked as a teacher at various points in his career. He was interested in theoretical problems, including the mathematically sophisticated construction of map projections and aided Gilbert in his philosophical enterprise. On the other hand, this was an academic who respected practical experience. He himself experienced the problems of ocean navigation, he built instruments, and he solicited the help and opinion of sailors and navigators. His motivation for this balancing of handwork and brainwork were many, probably including financial gain and social prestige as well as more intellectual concerns. He was certainly concerned with the usefulness of his investigations and, through the patronage support of aristocrats, Prince Henry, and the East India Company (somewhat latterly), was able to argue the utility of geographical knowledge both to imperial and mercantile causes.

Another preeminent figure in mathematical geography, also connected with Prince Henry was Thomas Harriot.[84] Harriot attended Oxford at the same time as Wright was at Cambridge. He matriculated from St. Mary's Hall in 1577 and received his B.A. there in 1580. By 1582 he was in the employ of Sir Walter Ralegh, who sent him to Virginia in 1585. Harriot, like Wright, was an academic

[79]"Mr. Sherburne's Appendix to his translation of Manilius, p. 86," in Birch, *Life of Henry, Prince of Wales*, 389.

[80]R. Malcolm Smuts, *Court Culture and the Origins of a Royalist Tradition in Early Stuart England* (Philadelphia: University of Pennsylvania Press, 1987). Smuts especially mentions Salomon de Caus, *La perspective avec la raison des ombres et miroirs* (London, 1612), dedicated "Au Serenissime Prince Henry," 157.

[81]Roy Strong, *Henry, Prince of Wales and England's Lost Renaissance* (London: Thames and Hudson, 1986), 218, and Edward Wright, *Plat of part of the way whereby a newe River may be brought from Uxbridge to St. James, Whitehall, Westminster, the Strand, St. Giles, Holbourne and London*. MS. 1610, identified by E.G.R. Taylor, *Late Tudor and Early Stuart Geography 1583–1650* (London: 1934), 235.

[82]Strong, *Henry, Prince of Wales*, 212.

[83]Waters, *Art of Navigation*, 320–1.

[84]J.W. Shirley, *Thomas Harriot: A Biography* (Oxford: Oxford University Press, 1983).

and theoretical geographer whose sojourn into the practical realm of travel and exploration helped form his conception of the vast globe and of what innovations were necessary to travel it. Harriot's description of Virginia, seen in his *Brief and true report of … Virginia* (1588),[85] was "the first broad assessment of the potential resources of North America as seen by an educated Englishman who had been there."[86] Harriot compiled the first word list of any North American Indian language (probably Algonquin),[87] a necessary first step of classifying in order to control, thus illustrating that inductive spirit never far from the heart of even the most mathematical geographer. He saw Virginia's great potential for English settlement, provided that the natives were treated with respect and that missionary zeal and English greed were kept to a minimum.[88] His advice concerning Virginian settlement was to prove important as the Virginia companies of the seventeenth century were established. This was the work of a man very aware of the practical and economic ramifications of the intellectual work of describing the larger world, as well as the imperial imperatives at work.

More important for Harriot were issues of the mathematical structure of the globe. Indeed his mathematics was bound up closely with his imperial attitude generally and the experience of his Virginian contacts in particular.[89] He was deeply concerned about astronomical and physical questions, including the imperfection of the moon and the refractive indexes of various materials.[90] Harriot was inspired by Galileo's telescopic observations of the moon and produced several fine sketches himself after *The Starry Messenger* appeared. He also investigated one of the most pressing problems of seventeenth-century mathematical geography – the

[85] Harriot, *Briefe and true report*, reproduced verbatim in T. de Bry, *America. Pars I*, published concurrently in English (also Frankfurt, 1590), and in Richard Hakluyt, *Principal Navigations* (1598), vol. 3, 266–280.

[86] David Beers Quinn, "Thomas Harriot and the New World," in Thomas Harriot. *Renaissance Scientist*, ed., J.W. Shirley, (Oxford: Oxford University Press, 1974), 45. See Amir R. Alexander, *Geometrical landscapes: the voyages of discovery and the transformation of mathematical practice* (Stanford: Stanford University Press, 2002) for an interesting interpretation of Harriot's mathematics.

[87] J.W. Shirley, *Thomas Harriot. Biography*, 133.

[88] The manuscript information concerning this expedition is gathered together in D.B. Quinn, *The Roanoke Voyages, 1584–1589: Documents to Illustrate the English Voyages to North America* (London: Hakluyt Society, 1955). J.W. Shirley discusses Harriot's desire for non-interference, *Thomas Harriot. Biography*, 152 ff. To see White's illustrations of this expedition, see "Picturing the New World. The Hand-Coloured De Bry Engravings of 1590," University Library, University of North Carolina at Chapel Hill, 2006, www.lib.unc.edu/dc/debry/about.html (Viewed December 9, 2007).

[89] Amir Alexander, "The Imperialist Space of Elizabethan Mathematics," *Studies in the History and Philosophy of Science* 26 (1995), 559–591, argued that Harriot's work on the continuum were influenced by his view of geographical boundaries and the 'other'. "The geographical space of the foreign coastline and the geometrical space of the continuum were both structured by the Elizabethan narrative of exploration and discovery," 591. Alexander develops this further in his book, *Geometrical Landscapes*.

[90] J.W. Shirley, *Thomas Harriot. Biography*, 381–416.

problem of determining longitude at sea. Harriot worked long and hard on the longitude question and on other navigational problems, relating informally to many mathematical geographers his conviction that compass variation contained the key to unraveling the longitude knot.[91]

Harriot was a mathematical tutor to Sir Walter Ralegh for much of the last two decades of the sixteenth century, advising his captains and navigators, as well as pursuing research interesting to Ralegh. As Richard Hakluyt said of Harriot, in a dedication to Ralegh:

> By your experience in navigation you saw clearly that our highest glory as an insular kingdom would be built up to its greatest splendor on the firm foundation of the mathematical sciences, and so for a long time you have nourished in your household, with a most liberal salary, a young man well trained in those studies, Thomas Hariot, so that under his guidance you might in spare hours learn those noble sciences.[92]

As Ralegh fell from favour, eventually ending up in the Tower, Harriot began to move his patronage expectation to another aristocrat interested in mathematical and geographical pursuits, the ninth Earl of Northumberland (the so-called "Wizard Earl"). Although Harriot's relationship with Northumberland is somewhat obscure, he appears to have conducted research within Northumberland's circle and occasionally household, as well as acting as a tutor as needed. Finally, Harriot was also connected with Henry, Prince of Wales, as a personal instructor in applied mathematics and geography, just as Wright was.[93] It is likely that Wright and Harriot met at Henry's court. As two university-trained contemporaries, with very similar interests and experiences, they would have gained much from their association. Given their mutual interests, it would have made sense for them to discuss matters of mutual geographical and mathematical interest while at court together.

Harriot's career displays many of the same characteristics as Wright's. Harriot too was a man who drifted in and out of academic pursuits, from university, to Virginia, to positions as researchers and tutors for Ralegh and Northumberland. In some ways, he was less connected to practical pursuits than Wright, although his trip to Virginia, and his work on longitude indicate his engagement with issues of practical significance. Harriot was also very dependent on patronage, especially that of Ralegh and of Northumberland (poor choices as they turned out to be), and used this patronage to help create an intellectual community where mathematical theory and imperial utility could be considered equally important.

Wright and Harriot, as well as a host of other geographers interested in this interconnection between theoretical and practical issues, combined an interest in the construction of the globe and a new more wide-reaching understanding of basic geographical concepts, with a desire for political and economic power on

---

[91] Harriot wrote a manuscript in 1596, entitled "Of the Manner to observe the Variation of the Compasse, or of the wires of the same, by the sonne's rising and setting," B.L. Add. MS 6788.

[92] Richard Hakluyt, introduction to Peter Martyr, as quoted in J.W. Shirley, "Science and Navigation," 80. See J.W. Shirley, *Thomas Harriot. A Biography.*

[93] J.W. Shirley, "Science and Navigation," 81.

the part of princes, nobles, and merchants. This wide-ranging area of investigation encouraged associations to develop between academic geographers, instrument-makers, navigators, and investors. The result was a negotiation between theoretical and practical issues, which also allowed a new negotiation with nature. This fruitful association between theory and practice helped to determine the kinds of questions these men asked, the kinds of answers that were acceptable, and the model of the world that would be developed. Thus, at least in this area of scientific interest, this socio-economic and political structure deeply influenced the development of what we would now call the scientific revolution.

2.7 Conclusion

Wright and Harriot provide a good example of the kind of investigators necessary for the development of the 'scientific revolution'. These two men, and many other mathematical practitioners, represent the communication between theory and practice, both within their own careers and ideas, and between universities, courts, print shops, the shops of instruments makers, and many other liminal venues. Their lives and careers show that new locales were becoming important for the pursuit of natural knowledge, including urban shops and houses on the one hand, and the courts and stately homes of aristocratic and noble patrons on the other. Wright and Harriot also demonstrate within their scientific worldviews an interesting mixture of theory, inductive fact-gathering, and quantification, which provided part of the changing view towards nature and its investigation so important for the 'new science'. They were both concerned with practicality and utility, especially within the rhetoric they employed to argue their cause, but also in the problems they tackled and the answers they thought sufficient. Finally, their connections to mercantilism are illustrative – extremely important, to be sure, but not consuming their lives. This was not science directed by the bottom line of mercantilist expenditure, but rather a more complex interaction between court, national and international intellectual community, and mercantilist enterprise.

Thus, the mathematical practitioners provided an agent for the changing nature of the scientific enterprise in the early modern period. They made the link in communication between theory and practice so fundamental for Zilsel. They did so for reasons that included the economic and bourgeois changes that were affecting Europe so directly, though not particularly with the rise of the towns. As we will see in the papers in this volume, these men were also concerned with issues of nationalism, imperialism, cultural credit, and status, issues that do not fit easily into Zilsel's more materialist interpretation.

Did they create the scientific revolution or a new science? In a sense, yes. Because these men were interested in mathematics, measurement and quantification became increasingly more significant. Their social circumstances ensured that the investigation of nature must be seen to be practical, using information from any available source, and science developed a rhetoric of utility and progress, as well as an inductive methodology, in response. Intimately connected to national pride and mercantile profit, the science that developed in this period reflected those concerns, a heritage modern science might like to forget. In essence, in large part because of the work of mathematical practitioners like Wright and Harriot, the investigation of nature began to take place away from the older university venue (though there remained important connections), with new methodologies, epistemologies, and ideologies of utility and progress. The scientific revolution had begun.

But there was still something missing. Wright and Harriot did not make the transition to natural philosophers. Despite their interest in patronage and the natural world, they remained mathematical practitioners. And by the end of the seventeenth century, mathematical practitioners had been reduced to technicians, whose presence became more and more invisible.[94] Meanwhile, natural philosophers, like Robert Boyle, or Isaac Newton, removed themselves from the company of mathematical practitioners, even as they used the fruit of their labour. This termination of the conversation between scholar and craftsman is just as important as its initiation; it is however the story of the legitimation of the scientific revolution, and therefore must be left to another day.

Zilsel was right to argue for the importance of economic developments and of the influence of technical and technological issues. He was also correct to see the importance of the mechanics' art to the mechanical philosophy, that is, the expertise and participation of artist-engineers (most particularly mathematical practitioners) in reconceptualizing the globe and the natural world more generally. What he missed was the mechanism. The changes of the sixteenth century were not just economic, but involved a fundamental change in politics, culture, and religion as well. Most importantly for the changes in investigations of the natural world – the scientific revolution – were the rise of new political units, with their focus on the courts, the rise of a new religious diversity which caused people to question the Catholic Church's earlier monopoly on truth, and the rise of a new group of men interested in advancing in the political world through hard work and political acumen. The result was that understanding of the natural world – now useful and applicable – was a new status symbol within this culture.

[94]Brotton shows that cosmographers had become employees of the joint stock companies by the end of the seventeenth century (*Mapping Territories*, 186), while Shapin, *Social History* argues for the increasing invisibility of technicians, 355–408. Thomas Sprat, *The History of the Royal Society of London* (London, 1667), 392, celebrates the distance between gentlemen who create new knowledge and technicians who can only do as they are told. Lisa Jardine, *The curious life of Robert Hooke: the man who measured London* (New York: Harper Collins, 2003), suggests that Hooke remained a technician.

Chapter 3
Consuming and Appropriating Practical Mathematics and the Mixed Mathematical Fields, or Being "Influenced" by Them: The Case of the Young Descartes

John A. Schuster

Abstract This chapter aims to clarify how historians can address the problem of what early modern practical and mixed mathematics had to do with the contemporary transformation of natural knowledge, taking the latter primarily as a set of changes in the domain of natural philosophy. It examines this problem from the perspective of natural philosophical consumers of resources—technical, theoretical and rhetorical—provided by mathematical practitioners and devotees of the mixed mathematical disciplines. The chapter criticizes historical narratives which speak of practical or mixed mathematics 'influencing' and 'shaping' natural philosophy, proposing that the relationship is better understood as a process of appropriating and translating resources between one field and another. Also questioned are prevalent narratives in which a 'target' (e.g. 'science') is influenced by a 'source' (e.g. practical mathematics) to produce some grand and essential change such as the 'birth of modern science'. Four case studies support this analysis: Three are drawn from the author's earlier studies of the young Descartes' aspirations in physico-mathematics and mechanistic natural philosophy; the fourth deals with the question of the appropriation and transformation of mechanics from practical and mixed mathematics into natural philosophy, in which Descartes played a part.

3.1 Externalist Narrative, the New Historians of Practical Mathematics and the Category of Natural Philosophy

Since I am not an historian of practical mathematics, I have no intention of adding to the substantive deliberations of the distinguished historians of practical mathematics brought together in this volume. Rather, as an historian of the Scientific

J.A. Schuster (✉)
Unit for History and Philosophy of Science, University of Sydney, Sydney, Australia
e-mail: drjaschuster@gmail.com

L.B. Cormack et al. (eds.), *Mathematical Practitioners and the Transformation of Natural Knowledge in Early Modern Europe*, Studies in History and Philosophy of Science 45, DOI 10.1007/978-3-319-49430-2_3

Revolution, my orientation is toward the general question of what role[s] practical mathematics played in the Scientific Revolution. My concerns reside with the culture and dynamics of what early modern actors called 'natural philosophy', as well as the specialist disciplines those actors held to be subordinate to natural philosophizing, especially the fields of mixed mathematics, such as geometrical astronomy, optics, statics, and music theory. Hence, I ask, *"What did practical mathematics and mathematicians have to do with changes in early modern natural philosophy and its subordinate disciplines, and why and how did this happen'?* And, I do this not by looking *in* from practical mathematics toward the Scientific Revolution, but rather looking *out* from the culture of natural philosophy to see how practical mathematics and its resources—technical, theoretical and rhetorical—were received and appropriated by innovative natural philosophers of the period.

Four case studies are presented, the first two of which examine Descartes' early work in hydrostatics and geometrical optics, and his appropriation of that work into the construction of his brand of mechanical philosophy. Attention is paid to the way he practiced the mixed mathematical sciences, which most Aristotelians held to be subordinate to natural philosophy in the sense of being of instrumental value only, and incapable of treating questions pertaining to matter and cause. Descartes tried to render the mixed mathematical fields more 'natural philosophical' in character—or, as he would have said in his early years, more 'physico-mathematical'. Since his view of practical mathematics was implicated in these developments, these two cases illuminate the young Descartes' transactions with practical mathematics in the service of what we may term 'the physicalization of the mixed mathematical sciences'. The paper also makes a number of historiographical suggestions regarding the explanation of the 'Scientific Revolution'; the relevance of the practical mathematics tradition to that problem; and the avoidance of pitfalls in approaching these issues. This is done in the more historiographical sections of the paper, as well as through two shorter case studies, dealing with natural philosophers' appropriation of the sixteenth-century mechanics tradition, and Descartes' complicated transactions regarding his lens grinding machine.

Before we examine those case studies or arrive at any new historiographical insights, we must first review our inherited starting point for thinking about practical mathematics and the Scientific Revolution. This, it turns out, is a special case of traditional externalist narrative of the Scientific Revolution. By unpacking the traditional externalist problematic, we shall be better placed to appreciate the approach I am advocating, whilst still perceiving its continuities with the older externalist impulse. When relating practical mathematics to the Scientific Revolution, historians of practical mathematics usually see mathematical practitioners as agents of change, and the object of that change being the method and ideology of science. The method becomes mathematical and instrumental, whilst the ideology values material practice and social utility. This is perfectly consistent with the problematic of traditional externalism in the historiography of science as promoted by Hessen, Zilsel, Needham and others. They variously argued that practical mathematics (often taken as part of a larger movement of the practical arts) had played the seminal role in the establishment of modern science according to the following externalist emplotment: Modern science, product of the Scientific Revolution, was a goal, possessing an essence, which consisted in mathematicized theory, proper method, and the values of utility and social progress. *Theory* meant correct, definitive theory of a mathematicized nature, in mechanism, Copernicanism or Newtonianism; proper *method* conjoined mathematics with experiment. Practical mathematics supplied DNA for that essence and was itself powered by new economic demands and technical problems arising there from.[1]

There are modern versions of this emplotment. In Paolo Rossi's compelling story, the practical arts *in general* play the lead. From the mid- to late sixteenth century the elite end of that literature expressed the values which Bacon and the early mechanists later implanted in high cultural natural philosophizing and precipitated the essence of the new science.[2] Similarly Jim Bennett,*doyen* of the new history of practical mathematics, has followed a similar emplotment on those occasions when he has provided a master narrative: Practical mathematics finally had its pay-off in the emergence of the mechanical philosophy, whose essence consists in experimental practice, instrument deployment, mathematical formulation and mechanistic explanation, all DNA borrowed from the practical mathematics.[3] In sum, the old externalism haunts our historical imaginations, threatening to materialize whenever we attempt big pictures of the relation of practical mathematics to the rise of modern science, so that, unless we are careful, we intone something that amounts to:

[practical mathematics] ➔ [causes/shapes] ➔ [modern science]

Now, since the business of this volume is to ask again, "What was the role of practical mathematics in the Scientific Revolution?", we need to think through our inherited externalist emplotment at a broad historiographical level, so that we can, at a general level, move beyond it.

---

[1]See John Schuster, "Internalist and Externalist Historiographies of the Scientific Revolution," in Wilbur Applebaum (ed.), *Encyclopedia of the Scientific Revolution* (New York, 2000), 334–6; John Schuster, "The Scientific Revolution," in Robert Olby, Geoffrey Cantor, John Christie and M.J.S. Hodge (eds.), *The Companion to the History of Modern Science* (London, 1990), 218–222; and Stephen Shapin, "Discipline and Bounding: The History and Sociology of Science As Seen Through the Externalism-Internalism Debate," *History of Science* 30 (1992): 333–369.

[2]Paolo Rossi, *Philosophy, Technology and Arts in Early Modern Europe* (New York, 1970).

[3]Jim Bennett, "The Mechanics' Philosophy and the Mechanical Philosophy," *History of Science* 24 (1986): 1–28; Bennett, "The Challenge of Practical Mathematics," in S. Pumfrey, P. *L. Rossi* and M. Slawinski (eds.), *Science, Culture and Popular Belief in Renaissance Europe* (Manchester, 1991), 176–190; and Bennett, "Practical Geometry and Operative Knowledge," *Configurations* 6.2 (1998): 195–222. There is more to Bennett's historiography, and we shall later return to his very fruitful, less mundanely externalist emplotments.

3.1.1 *Too Many Targets, Too Many Sources, Too Many Modes of Causation*

Externalist talk may be analyzed under categories I term 'source', 'mode of causation', and 'target', defined as follows:

Source Is it the practical arts in general, or some particular sector of the practical arts that affect change in the sixteenth century?: mechanics; practical mathematics (or some part thereof, such as geography, algebra, or instruments); or the rhetoric of men of practice, their social habituses and values?

In the literature on practical arts/practical mathematics and the Scientific Revolution, we find multiple sources for the same target: Geography supplies method, but so does algebra, or instrumental practice,[4] whilst for Needham it was the West's unique mixing of artisans, proto-methodologists, with scholars in need of a method fix.[5] Similarly, there are various sources accounting for the 'target', mechanical philosophy: For Rossi, it is the values and aims of practical artisans in general; for Bennett, the attitudes and modes of practice of practical mathematicians; for others it is sixteenth-century mechanics, or reflections on clockwork and/or automata.[6]

Mode of Causation In externalist narratives, we often encounter appeals to the causal concept of 'influence', despite correct calls for its demise over the last generation by Quentin Skinner and colleagues, as well as leading sociologists

[4]Lesley Cormack, "Geography", in Wilbur Applebaum (ed.), *Encyclopedia of the Scientific Revolution: from Copernicus to Newton* (New York, 2000), 261–264; David Livingston, "Geography," in Robert Olby, Geoffrey Cantor, John Christie and M.J.S. Hodge (eds.), *The Companion to the History of Modern Science* (London, 1990), 743–760; Michael S. Mahoney, "The Beginnings of Algebraic Thought in the Seventeenth Century," in Stephen Gaukroger (ed.), *Descartes: Philosophy, Mathematics and Physics* (Brighton, Sussex, 1980), 141–155.

[5]Joseph Needham, *The Great Titration: Science and Society East and West* (London, 1969), 49–50; similarly for 'method' as the target, see Edgar Zilsel, "The Sociological Roots of Science," *American Journal of Sociology* 47 (1942): 544–62; or Boris Hessen, "The Social and Economic Roots of Newton's "Principia", in *Science at the Crossroads, Papers Presented to the International Congress of the History of Science and Technology Held in London from June 29th to July 3rd, 1931, by the Delegates of the USSR* (London, 1931), 149–212; for natural law as the 'target' see Zilsel, "The Genesis of the Concept of Physical Law," *Philosophical Review* 51 (1942), 245–79; for Newtonian physics, see Hessen also; for the mechanical philosophy as 'target', see, for example, Franz Borkenau, *Der Ubergang vom feudalen zum burgerlichen Weltbild. Studien zur Geschichte der Manufakturperiode* (Paris, 1934).

[6]Rossi, *Philosophy, Technology and Arts in Early Modern Europe;* Bennett, "The Mechanics' Philosophy and the Mechanical Philosophy"; Helen Hattab, "From Mechanics to Mechanism: The *Quaestiones Mechanicae* and Descartes' Physics," in Peter Anstey and John Schuster (eds.), *The Science of Nature in the Seventeenth Century: Changing Patterns of Early Modern Natural Philosophy* (Dordrecht, 2005), 99–129. Dordrecht: Springer, 2005.; Derek J de Solla Price, "Automata and the Origins of Mechanism and the Mechanistic Philosophy," *Technology and Culture* 5 (1964): 9–42; Otto Mayr, *Authority, Liberty and Automatic Machinery in Early Modern Europe* (Baltimore, 1986).

of scientific knowledge.[7] In other species of externalism we meet either a kind of magical social structural imprinting upon the thoughts of cultural dopes like Descartes and Newton;[8] or, more convincingly, some kind of Zilselian causation via social proximity (which still leaves problems); or, more sophisticatedly still, in Biagiolian/Shapinian historiography, a displacement of social types: mathematicians (or experimenting gentlemen) replace/displace mere natural philosophers.[9]

Target What is the 'thing' being shaped, influenced, brought into existence—Science; Mechanical Philosophy, scientific method, or new scientific values?

Hence, there are problems across the board about target, source and mode of causation: We have multiple targets for the same source, and multiple sources for the same target, with little attempt to think through the modes in which the causes work, let alone consensus on how to approach them. Clearly, we need to eschew classical externalist talk, and to take stock of the multiplication of purported targets and sources. The way forward is through conceptual and historiographical house cleaning, and fortunately, the tools for this are at hand in other corners of the scholarship. To begin, we may learn from recent moves in another troubled area of Scientific Revolution historiography: the problem of science and religion. Margaret Osler has proposed replacing simplistic metaphors of conflict, separation and harmony with new metaphors of mutual appropriation and translation, designed to emphasize the interactions between theology and natural philosophy.[10] Accordingly, we should decide straightaway that talk of "influencing", or "shaping/imprinting" must go. We should think, rather, of people borrowing, adapting and appropriating; but borrowing, adapting and appropriating what? Well, obviously, material and discursive resources—and so the defining questions become, "Who were the borrowers and in what tradition, or field did they reside?" That is, if we get the 'target' group (*i.e.*, the active agents)[11] right, causal mode sorts itself out as appropriating and translating, and the appropriators themselves should reveal their sources.

[7]Quentin Skinner, "Meaning and Understanding in the History of Ideas," *History and Theory* 8 (1969):3–53; Jan Golinski, *Making Natural Knowledge: Constructivism and the History of Science* (Cambridge, 1998); Barry Barnes, *T.S.Kuhn and Social Science* (London, 1982).

[8]Mary Douglas, *Natural Symbols* (New York, 1970), see pp. 77–92 for the notorious group/grid theory which enjoyed a brief fad in historiography of science; David Bloor, *Knowledge and Social Imagery* (London, 1976).

[9]On the pitfalls of this last option, see John Schuster and Alan Taylor, "Blind Trust: The Gentlemanly Origins of Experimental Science," *Social Studies of Science* 27 (1997): 503–536.

[10]Margaret. Osler, "Mixing metaphors: Science and Religion or Natural Philosophy and Theology in Early Modern Science," *History of Science* 36 (1998): 91–113.

[11]There is no mistake here. Once we have corrected our explanatory categories, the natural philosophers who were the 'targets' of influence or imprinting stories become the agents in revised narratives, active appropriators and translators of cultural resources and artifacts.

3.1.2 Natural Philosophizing as Culture and Process

My key suggestion about the target group—the active agents—is that we must employ the category 'natural philosophy' in preference to Science, Modern Science, new science, or another term. 'Natural philosophy' is *the* appropriate historical category with which to think through our problem, because in the early modern period it was the central discipline for the study of nature.[12] Early modern natural philosophy was a dynamic, elite sub-culture and field of contestation. When one 'natural philosophized', one tried systematically to explain the nature of matter, the cosmological structuring of that matter, the principles of causation, and the methodology for acquiring or justifying such natural knowledge. (Fig. 3.1) The dominant genus of natural philosophy was Aristotelianism in various Neo-Scholastic species, but the term applied to alternatives of the various competing genera: neo-Platonic, Chemical, Magnetic, Hermetic, mechanistic or, later, Newtonian. Natural philosophers learned the rules of natural philosophizing at university whilst they studied the hegemonic Scholastic Aristotelianism. Because even alternative systems followed the rules of this game, all natural philosophers constituted one sub-culture in dynamic process over time.

We should therefore not identify natural philosophy with Scholastic Aristotelianism only; nor should we imagine that natural philosophy died and was rupturally replaced by an essentially different activity: Science. The 'Scientific Revolution' largely consisted in a set of transformations inside the seething, contested culture of natural philosophizing. Under internal contestation and external drivers natural

[12]To place the evolution of natural philosophy, and its shifting patterns of relations to other enterprises and disciplines, at the center of one's conception of the Scientific Revolution is not novel, but neither is it widely accepted in the scholarly community. Attempts to delineate the category of natural philosophy and deploy it in Scientific Revolution historiography include, Schuster, "The Scientific Revolution"; Schuster, "Descartes *Agonistes* New Tales of Cartesian Mechanism," *Perspectives on Science* 3 (1995): 99–145; John Schuster and Graeme Watchirs, "Natural Philosophy, Experiment and Discourse in the Eighteenth Century: Beyond the Kuhn/Bachelard Problematic," in Homer E. LeGrand (ed.), *Experimental Inquiries: Historical, Philosophical and Social Studies of Experiment* (Dordrecht, 1990), 1–48; Andrew Cunningham, "Getting the game Right: some Plain Words on the Identity and Invention of Science," *Studies in History and Philosophy of Science* 19 (1988): 365–89; Andrew Cunningham, "How the *Principia* Got its Name; or, Taking Natural Philosophy Seriously," *History of Science* 24 (1991): 377–92; Andrew Cunningham and Perty Williams, "De-centring the 'Big Picture': The Origins of Modern Science and the Modern Origins of Science," *British Journal for the History of Science* 26 (1993): 407–32; Peter Dear, "The Church and the New Philosophy" in Stephen Pumfrey, Paolo. *L. Rossi* and Maurice Slawinski (eds.), *Science, Culture and Popular Belief in Renaissance Europe* (Manchester, 1991), 119–139; Peter Dear, "Religion, Science and Natural Philosophy: Thoughts on Cunningham's Thesis," *Studies in History and Philosophy of Science* 32 (2001): 377–86; Peter Harrison, "The Influence of Cartesian Cosmology in England," in Stephen Gaukroger, John Schuster and John Sutton (eds.), *Descartes' Natural Philosophy* (London, 2000), 168–92; Peter Harrison, "Voluntarism and Early Modern science," *History of Science* 40 (2002): 63–89; Peter Harrison, "Physico-Theology and the Mixed Sciences: The Role of Theology in Early Modern Natural Philosophy," in Peter Anstey and John Schuster (eds), *The Science of Nature in the Seventeenth Century: Changing Patterns of Early Modern Natural Philosophy* (Dordrecht, 2005), 165–83; and John Henry, *The Scientific Revolution and the Origins of Modern Science*, 2nd ed. (Basingstoke, 2001).

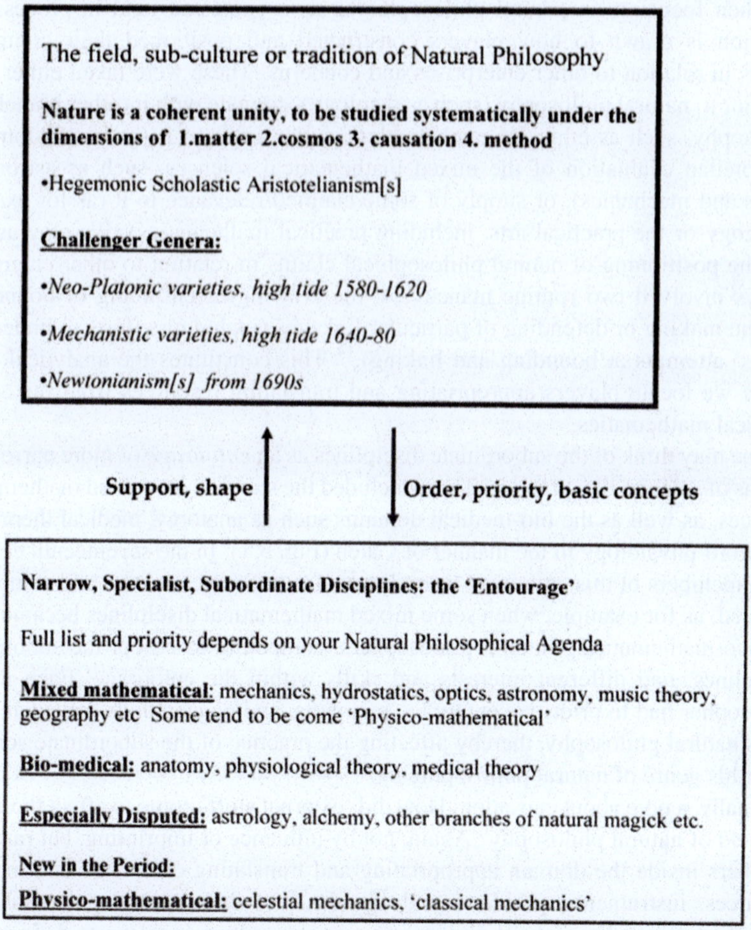

Fig. 3.1 Generic structure of natural philosophy and possible entourage of subordinate fields: In a given system of natural philosophy: (1) the particular entourage of subordinate disciplines lends support to and can even shape the system; while (2) the system determines the selection of and priority amongst entourage members, and imposes core concepts deployed within them

philosophy evolved, and eventually fragmented, into more modern looking, science-like, disciplines and domains over the period of approximately 150 years from 1650 to 1800.[13] This evolving complex is the 'target' in my 'source, mode and target' schema.

[13]John Schuster, "L'Aristotelismo e le sue Alternative," in Daniel Garber (ed.) *La Rivoluzione Scientifica* (Rome, 2002), 337–357; also Schuster and Watchirs, "Natural Philosophy, Experiment and Discourse in the Eighteenth Century"; and John Schuster, *Descartes–Agonistes*: *Physico–Mathematics, Method and Corpuscular–Mechanism, 1619–1633* (Dordrecht, 2013), 77–88.

When focusing on natural philosophizing as a contested field in process, our attention is drawn to how players constructed and positioned their competing claims in relation to other enterprises and concerns. These were taken either to be superior to natural philosophy (such as theology), cognate with it (other branches of philosophy, such as ethics or mathematics), subordinate to it (as in the dominant Aristotelian evaluation of the mixed mathematical sciences, such as astronomy, optics and mechanics), or simply of some claimed relevance to it (as for example pedagogy or the practical arts, including practical mathematics). We may assume that the positioning of natural philosophical claims in relation to other enterprises always involved two routine maneuvers: the drawing or enforcing of boundaries and the making or defending of particular linkages (including efforts to undermine others' attempts at bounding and linking).[14] This constitutes the analytical space where we locate players appropriating and translating resources from mixed and practical mathematics.

One may think of the subordinate disciplines as an *entourage* of more narrow traditions of science-like practice: These included the subordinate mixed mathematical sciences, as well as the bio-medical domains such as anatomy, medical theorizing, and proto-physiology in the manner of Galen (Fig. 3.1). In the seventeenth century, some members of this entourage were disputed, some were created, and some were changed, as for example, when some mixed mathematical disciplines became more physico-mathematical. Natural philosophers, competing to co-opt the subordinate disciplines, had different interests and skills within the entourage. Each natural philosopher had to prioritize entourage members, and conceptually articulate them to his natural philosophy, thereby affecting the practice of the subordinate sciences under his genre of natural philosophizing.

Finally, a note about causation: How did 'external stuff' come to affect the evolving field of natural philosophy? Again, not by influence or imprinting, but rather by members inside the domain appropriating and translating discursive and material resources, instruments, problems and agendas into their natural philosophizing. Thus, I conceptualize natural philosophy as a sub-culture in process, defined over time by the resultant of its players' combats over claims, where some of those claims involved responses to contextual forces, threats and opportunities. I see natural philosophical 'natives' (thinking along Marshall Sahlins' anthropological lines) adapting to challenges and opportunities by their own culturally specific moves, and not by being imprinted, influenced, or put out of business by 'Science'. Moreover, these moves were not determined by a universal logic and they could express considerable novelty, all the while remaining specific to the (evolving) culture.[15]

[14]Cf. Peter Anstey and John A. Schuster, "Introduction," in Peter Anstey and John Schuster (eds), *The Science of Nature in the Seventeenth Century*, 1–7; Schuster, *Descartes–Agonistes: Physico–Mathematics, Method and Corpuscular–Mechanism*, chapter 2.

[15]Attentive readers will note the debt my model owes to theoretical insights about cultural dynamics pioneered by the anthropologist Marshall Sahlins, "Goodbye to *Tristes Tropes:* Ethnography in the Context of Modern World History," *Journal of Modern History* 65 (1993): 1–25. He models cultures as dynamic historical entities, focusing on their mechanisms of adaptation to exogenous

I term this a *cultural process model* of the 'mode of causation'.[16] Returning to our theme, the 'role[s] of practical mathematics in the Scientific Revolution', we now have a way to envision the '*target*', natural philosophy, and the '*modes*' by which its mutually competing players (the natural philosophers) appropriated, translated and redeployed what they perceived as relevant and useful in one of the target's main external '*sources*', practical mathematics.

3.1.3 Practical Mathematics Was Also a Tradition in Process

We can now think through the relations of practical mathematics to natural philosophy, provided we realize that practical mathematics was also a changing and developing field, and hence that appropriation and translation occurred in both directions. For example, Jim Bennett has provided a number of partial definitions of practical mathematics as an internally complex, dynamic and contested field or tradition.[17] He further writes of a "domain of practical geometry", containing sub-domains such as practical astronomy, surveying, perspective, cartography, architecture, fortification, engineering and machines, the art of war, navigation, and dialing. The larger domain contained shared "disciplinary assumptions", "material resources and mathematical techniques,"[18] and "a recognised circle of practitioners and an understood, though expanding, domain of competence",[19] sharing a confidence in progress, and held

---

and endogenous challenges over time. He argues that cultures display specificity of response to outside impingement; they are not simply imprinted upon or pushed around. The dynamics of response, *over time*, characterizes the culture (*ibid.*, p. 25). Steven Shapin, "Discipline and Bounding: The History and Sociology of Science As Seen Through the Externalism-Internalism Debate," *History of Science* 30 (1992): 333–369, speaks in analogous ways of the various sciences as cultures in process.

[16]This model holds for all types of contextual drivers or causes of natural philosophy asserted by externalists. Not merely practical mathematics, but quite macro entities—social structure, economic forces, political structures and processes—can be appropriately brought into play. The arguably objective existence of contextual structures and processes that historians need to model and explain did not cause, imprint or 'influence' thoughts about natural philosophy by natural philosophers. Rather, natural philosophers responded to challenges and forces and decided to bring them into play in the form of revised claims, skills, material practices and values in the field. To do that, the 'things' being brought in had to be represented *to and by them (not us!)* in appropriate form.

[17]Bennett explicitly endorses the attempt to construct such a concept of practical mathematics and apply it to historical inquiry and explanation. Bennett, "Practical Geometry and Operative Knowledge", p. 198: "Comparative accounts of what geometers *do* in the fifteenth and sixteenth centuries reveal a recognized, though not static, domain of practice with shared disciplinary assumptions, which should inform and illuminate our historical narratives."

[18]Ibid.

[19]Bennett, "Practical Geometry and Operative Knowledge", p. 219.

together by a common legitimatory rhetoric.[20] Bennett's conception of this field or tradition focuses on groups sharing and developing particular instruments or geometrical techniques.[21] His and others' research shows that as individuals or groups pushed particular instruments, techniques and supporting rhetoric from one sub-domain to another, they tended to produce knock-on competitive effects. For example, just as navigation became for some a mathematical science, so elite practitioners attempted to push the trigonometry used in astronomy into surveying, leading to conflict between more theoretically oriented and more artisan-like practitioners.[22] Those commanding more sophisticated, theory-relevant techniques moved to displace more artisanal types from work and reputation. Bennett also stresses the "rhetorical" dimension of instruments, involving the self-image of the instrument's owner, the patron's status, the maker's ambition and his intended impression upon potential clients.[23]

Much more can be said about practical mathematics as a tradition in process; but, for present purposes, the following heuristic advice suffices.[24] First of all, one should focus on common artifacts, techniques, problem solutions and concepts, since these held the tradition together. The dynamics were then supplied by how tradition elements were transformed by players with different agendas, roles, access to patronage or other types of material support. Objectively determinable factors enter here: context specific distributions of university chairs; demands for types of instruction, and sites for their delivery; the distribution of sites for patronage; patterns of education and role expectation amongst the nobility, gentlemen and commercial classes.[25] Actions in the field depended both upon the perceptions of and the agendas regarding the capture of these resources and roles.

---

[20]The field of geography, as described by Bennett, serves as an early exemplar of a dynamism understood by mathematical practitioners and their audiences. Bennett, "Practical Geometry and Operative Knowledge", pp. 202–6. See also Walton, "Technologies of Pow(d)er" in this volume.

[21]An example of a shared tool kit is projective geometry, used in perspective painting, cartography, and instrument design. Bennett, "Practical Geometry and Operative Knowledge", pp. 198. For sixteenth century England similarly see Stephen Johnston, "Mathematical practitioners and instruments in Elizabethan England," *Annals of Science* 48 (1991): 319–344, and Stephen Johnston "The identity of the mathematical practitioner in sixteenth-century England," in Irmgard Hantsche (ed.), *Der 'mathematicus': Zur Entwicklung und Bedeutung einer neun Berufsgruppe in der Zeit Gerhard Mercators* (Bochum, 1996), 93–120.

[22]Bennett, "Practical Geometry and Operative Knowledge" pp. 206–7; and Bennett, "The Challenge of Practical Mathematics." pp. 179–81.

[23]Bennett, "Practical Geometry and Operative Knowledge", pp. 206–7. And as an example, see Aaron Rathborne's *The Surveyor* (1616).

[24]Material in this and the next paragraph arose through collaboration with Dr. Katherine Neal [Hill] and was first presented at the Quadrennial Joint HSS, BSHS and Canadian Society for the History and Philosophy of Science Conference, St Louis, Missouri, August 2000.

[25]For example, in the early seventeenth century, relatively centralized, monarchical France had fewer significant patronage sites than did Italy, but had many young gentlemen educated by the Jesuits, and therefore indoctrinated into the value of practical mathematics for the 'gentleman officer', destined for service in the religio-political conflicts of the time.

Returning to the relations between practical mathematics and natural philosophizing, it is clear these were characterized by mutual articulation, not one-way traffic, for practical mathematical work was occasionally affected by moves coming from natural philosophizing. For instance, Napier's development of the logarithms shows the importance of concepts of uniform and non-uniform acceleration and velocity to his approach. Moreover, his aim was astronomical, so some of his tools and aims arose from the domain of natural philosophizing. Similarly, as practitioners took parts of the mixed mathematical field of optics into the tradition of practical mathematics, their results in turn could be imported by natural philosophers and re-negotiated as part of their own trajectories in the natural philosophical contest.[26]

Two important insights emerge here: First, the simple (but multifarious) externalist stories of source, mode of cause, and target that invoke practical mathematics must be set aside in favor of the study of the mutual articulations and internal contestations over time in the trajectories of both traditions: natural philosophy *and* practical mathematics. Second, our modeling of both fields supports our earlier surmise that some of the most important action involving innovating natural philosophers and the realm of practical mathematics took place in the domain of mixed mathematics, which, according to the dominant Neo-Scholastic Aristotelianism of the universities, was ambiguously placed and subordinate to, but not organically part of, natural philosophy. Accordingly, we next use our new models to 'rectify' the old domain of externalist explanation, in preparation for our case studies.

3.2 Rectifying the Terrain of Externalist Explanation

In this section we examine mixed mathematics as a contested borderland between natural philosophy and practical mathematics. We also reconsider practical mathematicians' rhetoric concerning the utility and progressiveness of their domain, recalibrating how this element enters into revised narratives of 'practical mathematics and natural philosophizing'.

We begin with the question of the status of the mixed mathematical sciences according to the dominant Scholastic Aristotelianism: Natural philosophy studies matter and cause and renders physical explanations. Mathematics deals with geometrical figures and numbers—things that do not change and exist only in our

[26]See on this Sven Dupré, "The Making of Practical Optics," in this volume. My point here was initially stimulated by Jim Bennett's discussion of three cases of natural philosophical appropriation of practical mathematical resources—Tycho Brahe (practical astronomy), Leonard Digges (gunnery) and William Gilbert (navigational magnetism). Bennett, "Practical Geometry and Operative Knowledge", p. 220.

minds. On this basis Aristotelians recognized the so-called 'mixed' mathematical sciences, such as planetary astronomy, geometrical optics, statics, music theory and mathematical geography, as subordinate to natural philosophy. They can give only instrumental mathematical descriptions, not causal explanations. For example, according to Aristotelians, the investigation of the physical nature of light falls under natural philosophy, involving principles of matter and cause. The mixed mathematical science of geometrical optics is subordinate to both natural philosophy and mathematics. Studying ray diagrams, where geometrical lines represent rays of light, it deals with phenomena such as the reflection and refraction of light in a descriptive, mathematical manner, and cannot provide causal explanations, based on the physical nature of light. Such was the dominant, "declaratory" neo–Scholastic view of how the mixed mathematical disciplines related to the 'superior' discipline of natural philosophy.[27] Subsequent debates started from this hegemonic base.

One of the most attractive recent lines of inquiry looks to progressive Scholastics themselves, especially leading Jesuit mathematicians, for the decisive moves to liberate and more fully mathematize these sciences. Peter Dear wove a sophisticated narrative along these lines, focusing upon previously neglected Scholastic mathematicians: Early in the seventeenth century some "Jesuit mathematical scientists"—astronomers and opticians—began to attempt "to justify these disciplines against criticism of their scientific status".[28] Their strategic location in Jesuit colleges and universities amplified the import of these moves. Dear expertly followed a series of textbooks and debates amongst this group, which initiated the elaboration of a new, non-Aristotelian concept of singular, contrived and mathematically articulated 'experience'. This represented a bid for the disciplinary autonomy of the mixed mathematical sciences from 'natural philosophy'. Dear argued correctly that for Jesuit mathematicians, such as Clavius, "Mathematical sciences that applied to the physical world were not taken to be in conflict with qualitative Aristotelian natural philosophy, but were typically seen as being about different things."[29] Clavius and others used this mathematics/natural philosophy distinction to preserve the integrity and certitude of mathematical pursuits, hence to legitimate the mixed mathematical disciplines as of explanatory and scientific status. Dear says this demarcation enhanced "their own pretensions to scientificity, and set the stage 'for a co-optation of natural philosophy itself'"—the emergence of what Dear and his subjects termed "physico-mathematics".[30] This then fed into Dear's larger

[27] I term the widely taught rule of subordination of mixed mathematics to natural philosophy 'declaratory' to denote that it was publicly proclaimed, but not necessarily binding or agreed to by relevant players. See Schuster, *Descartes–Agonistes*: *Physico–Mathematics, Method and Corpuscular–Mechanism, 1619–1633*, p. 51.

[28] Peter Dear, *Discipline and Experience: The Mathematical Way in the Scientific Revolution* (Chicago, 1995), 6.

[29] Dear, *Discipline and Experience*, p. 163.

[30] Dear, *Discipline and Experience*, p. 150.

story of the rise of modern (mathematico-experimental) science. Others, including Mersenne, Descartes and Beeckman, developed physico-mathematics, and further mid-century developments eventually led to Newton, who perfected the needed extra ingredient of the one-off 'event experiment' to arrive at the "spiritual core of modern science".[31]

Dear's elegant account has one unfortunate and unintended undertone, in that it resembles an origin tale: Embryonic modern science was hived off from 'natural philosophy' (equated with Scholastic Aristotelianism only), which conveniently died. The difficulty is that the key figures in the early physicalization of the mixed mathematical sciences were not the Jesuit Aristotelian mathematicians, but the usual suspects in Scientific Revolution historiography, such as Galileo, Kepler, Descartes, Gilbert, Mersenne and Beeckman. Early in the seventeenth century, it was these natural philosophers who variously claimed that mathematics could play an explanatory role *in* natural philosophy, rejecting the declaratory Aristotelian position.[32] Moving between mixed mathematics and novel natural philosophizing, they produced more 'physico-mathematical' versions of the old fields, supportive of their respective natural philosophical agendas. The origin of mathematized sciences, is really the emergence of more physicalized versions of the existing mixed mathematical sciences, and the construction of some new ones—all within the bubbling field of natural philosophizing, as innovative natural philosophers competed to appropriate resources, technical and rhetorical, from a rich and dynamic practical mathematics tradition. All this serves to articulate the view of natural philosophy as a contested field in which players first learned the rules of claim-making through their Neo-Scholastic Aristotelian educations, but could realize that these rules were 'negotiable', as the sociologists of scientific knowledge would say, and that their meanings were in the hands of successive waves of users. While some Aristotelians tried to bend the rules about the subordinate nature of the mixed mathematical disciplines, more radical natural philosophers, such as Kepler and Descartes, often ran right over them, forging new meanings and practices. By the first third of the seventeenth century, the given rules of subordination of mixed mathematics were the subject of vexed debate. To bring resources from practical

[31]This is not meant as a full summary of Dear's widely appreciated argument. We are interested here in the earliest stages of the story: [1] the tactics of the Jesuit mathematicians, and [2] the wider spectrum of meanings of physico-mathematics at the time.

[32]See Schuster and Taylor, "Seized by the Spirit of Modern Science", *Metascience* ns 9 (1996): 9–28. We hold that the plays of Clavius and his colleagues were moves within the wider field of natural philosophizing, and somewhat precious and unproductive ones. Moreover, theirs was not the only version of physico-mathematics on offer, as we learn below. See also Schuster, *Descartes–Agonistes*: *Physico–Mathematics, Method and Corpuscular–Mechanism*, pp. 56–59 and John Schuster, "Physico–mathematics and the Search for Causes in Descartes' Optics—1619-37," *Synthèse* 185 (2012): 467–499.

mathematics into this arena was, to radical players, a very attractive gambit.[33] We shall touch on some technical matters later in our case studies. For the moment we concentrate on the rhetorical transactions involved.

Natural philosophical radicals, such as Descartes, Beeckman, Kepler, and Galileo, who were physicalizing the mixed mathematical sciences, operated within a discursive framing of their enterprises, based on an already available rhetoric of the utility, intelligibility and cognitive value of the mechanical arts and practical mathematics.[34] Masters of the practical arts, including practical mathematicians, had spent a lot of time in the sixteenth century publicizing the usefulness, and the knowledge-like character, of their enterprises. Our early seventeenth-century natural philosophers picked up these messages, reformatted them for natural philosophical utterance and rebroadcast them as legitimations for new agendas in natural philosophy.[35] Such co-options were endemic, and perhaps cumulative; we find them all along the trajectory of interactions. Consequently, one certainly should not mistake any instance of a natural philosopher co-opting the rhetoric of the practical mathematicians—a pitfall for the early externalists—for the 'foundation of the essence' of 'modern mathematical science'. Nevertheless, appropriation of practitioner's rhetoric was substantively important for innovative natural philosophers. It helped shape their self-understandings of their programs and it softened up audiences for their reception. To explain this further: As our case studies will show, these particular technical developments by Descartes had as their necessary causes technical resources and skills. In general, rhetorical resources cannot in themselves constitute or explain such achievements. However, as I have often emphasized in my analyses of the political and rhetorical roles of method doctrines in the life of the sciences, rhetorical transactions should be studied and woven into dense accounts of scientific and natural philosophical gambits.[36]

[33]See John Schuster, "What Was the Relation of Baroque Culture to the Trajectory of Early Modern Natural Philosophy," in Ofer Gal and R. Chen-Morris (eds.), *Science in the Age of Baroque*. [*Archives internationales d'histoire des idées* 208 (2013)], 13–45, at pp. 16–19, 21–28. Kepler, however, still paid non trivial 'declaratory' allegiance to them in some contexts. Cf. Rhonda Martens, *Kepler's Philosophy and the New Philosophy* (Princeton, 2000), Chapter 5 "The Aristotelian Kepler".

[34]Rossi, *Philosophy, Technology and Arts in Early Modern Europe*, has by far the best grasp of this process.

[35]Contemporary historians of practical mathematics, such as Jim Bennett, Catherine Neal [Hill], Stephen Johnson and Lesley Cormack, teach us that much conflict characterized the practical mathematical field. The common legitimatory 'front' about the value of the practical arts trumpeted by some natural philosophers, may therefore have had more to do with the natural philosophical *agon* than with any consensus amongst master mathematical practitioners. See, for example, Catherine Neal [Hill], "'Juglers or Schollers?': Negotiating the Role of a Mathematical Practitioner," *British Journal for the History of Science* 31 (1998): 253–274; and Catherine Neal [Hill],"The Rhetoric of Utility: Avoiding Occult Associations for Mathematics Through Profitability and Pleasure," *History of Science* 37 (1999): 151–178.

[36]My several previous publications on the mythopoeic character of general method discourses, Descartes' included, are synthesized in Schuster, *Descartes-Agonistes: Physico–Mathematics, Method and Corpuscular–Mechanism*, 8–10, 70–77 and 265–303. Such discourses cannot explain

Because the mixed mathematical sciences formed a borderland between natural philosophizing and the field of practical mathematics, one can "map" how mixed mathematics sat in relation to radical, anti-Aristotelian natural philosophizing and to practical mathematics. One can envision a spectrum of players: [a] natural philosophers little concerned about mixed or practical mathematics; [b] natural philosophers actively interested in co-opting and using technical and rhetorical resources from mixed and/or practical mathematics; [c] elite practical mathematicians abstracting from lower level practical mathematics who might or might not link their activities to natural philosophizing; and [d] lower level mathematical practitioners. The interesting action was in categories [b] and [c], where the physicalization of the mixed mathematical sciences took place, and new physico-mathematical disciplines emerged.

For mathematically inclined, anti-Aristotelian natural philosophers, such as Kepler, Galileo, Beeckman and Descartes, the mixed mathematical sciences were ripe for co-optation into their innovative natural philosophical pursuits. For such players, practical mathematics tagged along with the outcome for mixed mathematical sciences. For example, geometrical optics was involved in a wide range of mathematical practices and artifacts, whilst it also articulated with high level natural philosophical theorizing, and for some players, such as Kepler and Descartes, 'physicalized' versions of geometrical optics were a prime location for natural philosophical initiatives. Similarly, 'high' statics/hydrostatics was thought to ground understandings of simple machines, and through them, complex machines, and hence by extension, much of the 'mechanical arts'. This 'cultural fact' could be played upon from different directions by natural philosophers and practical mathematicians. For example, the young Galileo in his *de Motu* treated statics and hydrostatics dynamically in an (unsuccessful) attempt to extract from them anti-Aristotelian conclusions about natural and violent motion.[37] The view of the young Descartes, as we shall see, was that any rigorous result in the mixed mathematical sciences bespeaks the discovery of a deep physical truth, which can be reduced to corpuscular-mechanical terms. But, the mixed mathematical borderland could be appropriated in the other direction. The great Simon Stevin determinedly

---

technical achievements in the sciences, but actors routinely play appeals to such 'method-talk' into the accounts they render of their own and opponents' work as part of the continual process of making and breaking claims in the expert *fora* of scientific practice. Thus method-talk is indeed woven into the expert life of the sciences, but not in the manners in which believers in such methods attribute to them. On this latter point in particular see ibid., 293–297.

[37] Stephen Gaukroger, "The Foundational Role of Hydrostatics and Statics in Descartes' Natural Philosophy," in Stephen Gaukroger, John Schuster and John Sutton (eds.), *Descartes' Natural Philosophy* (London, 2000), 60–80; and Stephen Gaukroger and John Schuster, "The Hydrostatic Paradox and the Origins of Cartesian Dynamics," *Studies in History and Philosophy of Science* 33 (2002): 535–572.

	conservative	radical
Active in mixed maths	***Jesuit mathematicians, Simon Stevin (ultra conservative)***	***Kepler, Descartes, Harriot, and Galileo (but not a systematic natural philosopher)***
Inactive in mixed maths	***Garden variety Scholastic Aristotelians***	***Some involved in making classifications of mixed sciences, especially mechanics, as dealing with matter and cause; Gilbert?***

Fig. 3.2 View of relation of mixed/practical mathematics to natural philosophy. A classification of people talking about or practicing the mixed mathematical sciences

removed mixed mathematics and mathematical practice from the domain of natural philosophy (by which he understood Aristotelianism).[38]

Let us look more closely at conservative versus radical takes on the relation of mixed mathematics to natural philosophy, distinguishing those active in mixed mathematics from those not active, or merely talking about their classification and hence involved in rhetorical transactions *only* with practical mathematics (Fig. 3.2). The latter, in the lower right hand quadrant, includes those sixteenth century Scholastic commentators on the status of mechanics who edged toward acknowledging its relevance to natural philosophy, but who did not technically practice mechanics. In the upper left-hand quadrant are Dear's Jesuit mathematicians, active in mixed mathematics but holding a conservative view of their relation to natural philosophy —disciplinarily separate but scientifically 'equal'. Joining them is Stevin, master of the mixed mathematical fields, holding a different conservative view of radical separation, and mutual irrelevance. Ordinary Aristotelians, adhering to the declaratory subordination rule, occupy the lower left-hand quadrant, whilst the 'usual suspects', radical natural philosophers seeking to "physicalize" the mixed mathematical fields, are in the upper right-hand quadrant.

[38] Stevin endeavored to bring statics and hydrostatics, and the practices that follow from them, into an Archimedean, rigorous, mathematical context, thus rejecting the pseudo-Aristotelian *Mechanical Questions* with its dynamical approach to simple machines and statics. On Stevin see our first case study below and Note 47.

	[2] YES	[2] NO
[1] YES	**Galileo Harriot**	**Stevin and other elite mathematical practitioners**
[1] NO	**Can we identify any? Gilbert?**	**Most garden variety practitioners**

Fig. 3.3 Elite mathematical practitioners' agendas [1] synthesize practical and mixed mathematics beyond traditional understandings: yes/no [2] agenda articulated to the field of natural philosophy: yes/no

Next, let us ask of elite mathematical practitioners whether they tried to synthesize practical and mixed mathematics in any way beyond traditional understandings; and whether they linked such agendas to the field of natural philosophizing (Fig. 3.3). Expert mathematical practitioners, such as Galileo and Harriot, pushed practical and mixed mathematics beyond traditional understandings to extract natural philosophical capital. Stevin, pursuing higher cultural status for mixed and practical mathematics, but also denying their relevance to natural philosophizing, occupies the upper right hand quadrant. He is joined by advocates of the high status of practical mathematics, elite practitioners who, did not encroach into the domain of natural philosophy. Ordinary practitioners would be in the lower right-hand quadrant. The lower left-hand quadrant is reserved for those whose rhetoric or technical practice linked practical and mixed mathematics to natural philosophizing, but who had little impact on contemporary practices or understandings of the mathematical fields. One inhabitant might be William Gilbert, an innovative natural philosopher and a consumer of others' mixed and practical mathematical work, but not a particularly innovative practitioner therein.

In sum, we shall achieve better accounts of 'practical mathematics and the Scientific Revolution' if we think through the categories 'natural philosophy' and 'practical mathematics' in the ways suggested, and then follow the plays from each side into the 'marcher fiefdoms' of mixed mathematics. Therefore, in the latter portions of this study we shall turn to four case studies intended to illustrate and test these claims. But, in order to understand the first two cases, it is first necessary to pause to consider some concepts central to Descartes' mature natural philosophy.

3.2.1 *What Was Cartesian 'Dynamics', the Causal Register of His Natural Philosophy?*

In the first two case studies we shall be dealing with episodes in which the young Descartes, pursuing his physico-mathematical agenda, attempted to formulate the causal register of his mechanical philosophy, his dynamics, by appropriation and translation from certain domains of the mixed mathematical sciences. In order to understand these two cases, we first need to examine what Stephen Gaukroger and I term Descartes' "dynamics", a set of concepts that supplied the doctrine of physical causation within Descartes' natural philosophy.[39] As already noted, a core aim of 'natural philosophizing' was the identification of what causes material bodies to behave in particular ways. For example, in Aristotelianism, natural processes were explained primarily on the basis of causes identified with the nature or essence of the substance in question, while in neo-Platonic natural philosophies, brute matter was worked upon from the outside by various types of non-material causal agents. To theorize about matter and an associated 'causal register' was central to any genre of natural philosophy. Whatever disputes there might have been amongst Platonists, Aristotelians, Stoics, and atomists, there was consensus on what kind of theory provided the ultimate explanation of macroscopic physical phenomena, namely a theory of matter and causation. Descartes' mature natural philosophy was no exception, being concerned with the nature and 'mechanical' properties of microscopic corpuscles and a causal discourse, consisting of a theory of motion and impact, explicated through such key concepts as the 'force of motion' and 'tendencies to motion'. It is this causal register within Descartes' natural philosophy which we term his 'dynamics'.

In Descartes' mature corpuscular-mechanical natural philosophy, his carefully articulated theory of dynamics governed the behavior of micro-particles. Bodies in motion, or tending to motion, are characterized from moment to moment by the possession of two sorts of dynamical quantity: (1) the absolute quantity of the 'force of motion', which is itself conserved in the universe according to *Le Monde's* first rule of nature; and (2) the directional modes of that quantity of force, which Descartes termed 'determinations', introduced in *Le Monde's* third rule of nature. Descartes' dynamics focused on instantaneous tendencies to motion, rather than finite translations in space and time. As corpuscles undergo instantaneous collisions with each other, their quantities of force of motion and determinations are adjusted according to certain universal laws of nature, rules of collision.

[39] Gaukroger and Schuster, "The Hydrostatic Paradox and the Origins of Cartesian Dynamics", 557, 561, 568–70; John Schuster, "'Waterworld': Descartes' Vortical Celestial Mechanics: A Gambit in the Natural Philosophical Contest of the Early Seventeenth Century," in Peter Anstey and John Schuster (eds.) *The Science of Nature in the Seventeenth Century: Changing Patterns of Early Modern Natural Philosophy* (Dordrecht, 2005), 35–79, at pp. 38–41.

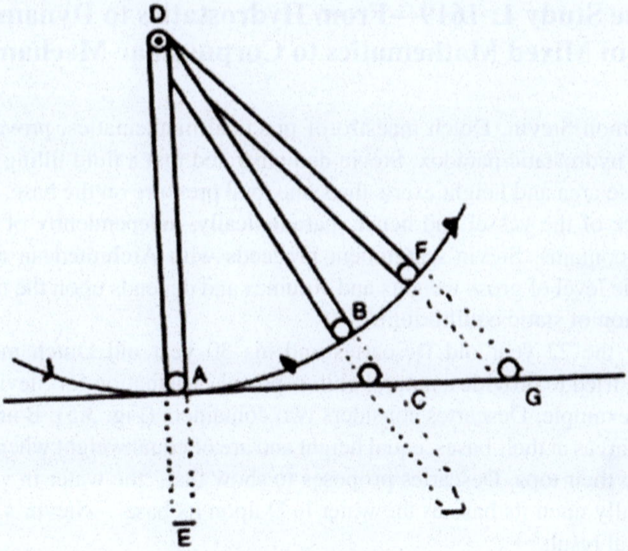

Fig. 3.4 After Descartes, Le Monde, AT XI p.45 and p.85

Descartes' exemplar for applying these concepts to light and celestial mechanics is the mechanics of a stone rotated in a sling.[40] (Fig. 3.4) He analyses the dynamical condition of the stone at the precise instant that it passes point A. The instantaneously exerted force of motion of the stone is directed along the tangent AG. If the stone were released and no other hindrances affected its trajectory, it would move along ACG at a uniform speed reflective of the moment-to-moment conservation of its quantity of force of motion. However, the sling continuously constrains the privileged, principal determination of the stone and, acting over time, deflects its motion along the circle AF. Descartes decomposes the principal determination into two components: one along AE completely opposed by the sling—so no actual centrifugal translation can occur—only a tendency to centrifugal motion; the other, he says, is "that part of the tendency along AC which the sling does not hinder", which over time manifests itself as translation in a circle. The choice of components of determination is dictated by the configuration of mechanical constraints on the system.

[40]Descartes, *Oeuvres de Descartes*. Edited by Charles Adam and Paul Tannery, 11 vols. (Paris, 1996) vol. XI pp. 45–6, 85 [Hereafter cited as AT (for Adam and Tannery edition, roman numeral for volume, plus page.]; Descartes, *Descartes, The World and Other Writings*. Edited and Translated by Stephen Gaukroger (Cambridge, 1998), 30, 54–5.

3.3 Case Study 1: 1619—From Hydrostatics to Dynamics: From Mixed Mathematics to Corpuscular Mechanism[41]

In 1586 Simon Stevin, Dutch maestro of practical mathematics, proved a special case of the hydrostatic paradox. Stevin demonstrated that a fluid filling two vessels of equal base area and height exerts the same total pressure on the base, irrespective of the shape of the vessel and hence, paradoxically, independently of the amount of fluid it contains. Stevin's argument proceeds with Archimedean rigor on the macroscopic level of gross weights and volumes and depends upon the maintenance of a condition of static equilibrium.[42]

In 1619 the 22 year old Descartes and his 30 year old Dutch mentor, Isaac Beeckman, tried to provide a natural philosophical explanation for Stevin's result.[43] In the key example, Descartes considers two containers (Fig. 3.5): B and D, which have equal areas at their bases, equal height and are of equal weight when empty, and are filled to their tops. Descartes proposes to show that, 'the water in vessel B will weigh equally upon its base as the water in D upon its base'—Stevin's paradoxical hydrostatical result.

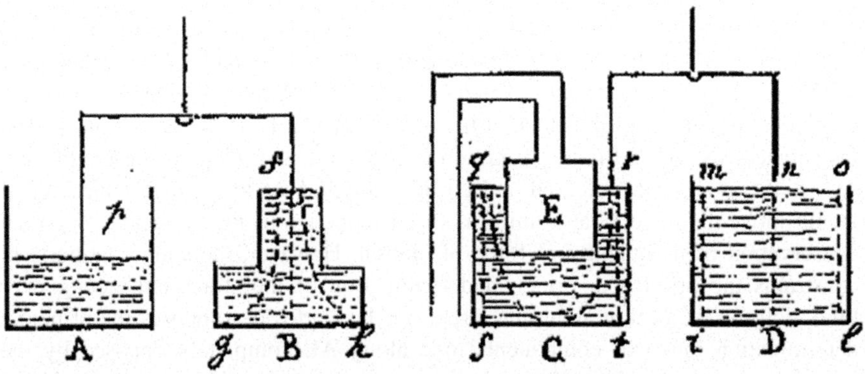

Fig. 3.5 Descartes, Aquae comprimentis in vase ratio reddita à D. DesCartes, AT X 69

[41]Material in this case derives from Gaukroger and Schuster, "The Hydrostatic Paradox and the Origins of Cartesian Dynamics"; Gaukroger, *Descartes: An Intellectual Biography* (Oxford, 1995), 84–9*;* and John Schuster, "Descartes' *Mathesis Universalis*, 1619–28," in Stephen Gaukroger (ed.) *Descartes: Philosophy, Mathematics and Physics* (Brighton, Sussex, 1980) 41–96 at pp. 41–55.

[42]Simon Stevin, "*De Beghinselen des Waterwichts*" (Leiden, 1586) in E, Cronie *et al.* (eds.), *The Principal Works of Simon Stevin.* 5 Vols. (Amsterdam,1955–66), Vol. 1, pp. 415–17.

[43]The text, *Aquae comprimentis in vase ratio reddita à D. Des Cartes* which derives from Isaac Beeckman's diary, is given in AT, X, pp. 67–74, as the first part of the *Physico-Mathematica.* See also the related manuscript in the *Cogitationes Privatæ*, AT, X, p. 228, introduced with, 'Petijt è Stevino Isaacus Middelburgensis *quomodo aqua gravitet in fundo vasis* b . . . '.

Descartes attempts to reduce the phenomenon to micro-mechanics by showing that the force on each 'point' or part of the bottoms of the basins B and D is equal, so that the total force is equal over the two equal areas. He claims that each 'point' on the bottom of B is serviced by a unique line of 'tendency to motion' propagated by contact pressure from a point (particle) on the surface of the water through the intervening particles. He takes points g, B, h; in the base of B, and points i, D, l, in the base of D. He cleverly claims that all these points are pressed by an equal force, because they are each pressed by 'imaginable lines of water of the same length'; that is, the same vertical component of descent. Despite this, Descartes' overall effort is distinctly odd. For example the mappings of lines of tendency are tendentious and not subject to any rule. Even so, for the rest of his career, Descartes continued to use descendants of these concepts of instantaneously exerted force of motion and its analysis into component 'determinations'.[44]

Descartes' manuscript signals that he no longer viewed hydrostatics as a discipline of mixed mathematics in the Aristotelian sense. Rather, he saw it as a domain of application of corpuscular-mechanical natural philosophy, because in order to explain the key hydrostatical results, concepts of matter and cause of clear natural philosophical provenance had to be deployed. This anti-Aristotelian program Descartes termed 'physico-mathematics',[45] but for several reasons it was a far cry from the physico-mathematics of Dear's Jesuit Aristotelian mathematicians: First, mixed mathematical hydrostatics is not severed from natural philosophy in order to secure it 'scientific status'; rather, it becomes coextensive with natural philosophical issues of matter and cause. Second, the species of natural philosophizing in question no longer is Neo-Scholastic Aristotelianism, but proto-mechanism. Finally, Descartes' approach was extremely radical, even within the small club of anti-Aristotelian physico-mathematical aspirants, because it was based in the rigorous style of Stevinite/Archimedean statics and hydrostatics, whereas most attempts to make anti-Aristotelian natural philosophical capital out of the mixed mathematical sciences depended on taking a *dynamical* approach to statics and the simple machines, following the lead of the pseudo-Aristotelian *Mechanical Questions*.

The *Mechanical Questions* views equilibrium conditions on a lever or simple machine as a balance of forces, where *force* is defined as *weight* times *speed.* Equilibrium is a special case of the dynamic opposition of the bodies; and statics

[44]Gaukroger and Schuster, "The Hydrostatic Paradox and the Origins of Cartesian Dynamics"; Schuster, "'Waterworld': Descartes' Vortical Celestial Mechanics"; John Schuster, "Descartes *Opticien:* Descartes' Manufacture of the Law of Refraction and Construction of its Physical and Methodological Rationales 1618–1628," in Stephen Gaukroger, John Schuster and John Sutton (eds.), *Descartes' Natural Philosophy* (London, 2000), 258–312.

[45]Descartes' employed the term physico-mathematics following lead of Beeckman, "Physico-mathematici paucissimi": AT X. 52. They clearly prided themselves on being virtually only true physico-mathematicians in Europe. In this regard Beeckman was later to note in 1628 that his own work was deeper than that of Bacon on the one hand and Stevin on the other just for this very reason. Isaac Beeckman, *Journal tenu par Isaac Beeckman de 1604 à 1634.* 4 vols. C. de Waard (ed.) (The Hague, 1939–53), Vol. 3, pp. 51–2.

is simply a limiting case of a general dynamical theory of motion.[46] Stevin, Descartes' exemplar in this matter, had preferred pure Archimedean statics and so had rejected this approach: dynamical thinking could not explain systems in static equilibrium.[47] However, Stevin had been in a minority on this issue[48] and the young Descartes daringly followed him, starting from a mathematically rigorous hydrostatics and then fleshing it out with Beeckman-esque corpuscles. The young Descartes' radical version of physico-mathematics involved reducing Stevin's hydrostatics to an embryonic corpuscular mechanism in which discourse concerning 'forces or tendencies to motion' would provide the basis for unifying the mathematical sciences, under a dynamics of corpuscles. His astonishing strategy was to appropriate practical and mixed mathematical materials, and creatively rework them through moves in the culture of natural philosophizing.[49] To confirm this, let us consider Descartes' work on refraction and physical optics, which, I contend, was the climax of his early physico-mathematical program.[50]

3.4 Case Study 2: 1627—The Laws of Light and the Laws of Nature[51]

The physico-mathematical hydrostatics of 1619 marked the first partial articulation of the central tenets of Descartes' dynamics. Their path of development between 1619 and the completion of *Le Monde* in 1633 led not through hydrostatics, but

[46]On the *Mechanical Questions* in this connection, see Gaukroger and Schuster, "The Hydrostatic Paradox", pp. 544 note 19. More generally, see Henri Carteron, *La Notion de force dans la système d'Aristote* (Paris, 1923); Pierre Duhem, *Les origines de la statique*, 2 Vols (Paris, 1905–6); Paul Lawrence Rose and Stillman Drake, "The Pseudo-Aristotelian *Questions of Mechanics* in Renaissance Culture," *Studies in the Renaissance* 18 (1971): 65–104; W.R. Laird, "The Scope of Renaissance Mechanics," *Osiris* 2 (1986): 43–68; and Helen Hattab, "From Mechanics to Mechanism".

[47]Gaukroger and Schuster, "The Hydrostatic Paradox", pp. 540, 545–9; Stevin, "Appendix to the *Art of Weighing"* in *Principal Works* Vol. 1, 507–9; and "*The Practice of Weighing*, 'To the Reader'", *ibid.* Vol. 1, 297.

[48]For example, the young Galileo had tried, unsuccessfully, to use the *Mechanical Questions* to found an anti-Aristotelian physics. Stephen Gaukroger, "The Foundational Role of Hydrostatics and Statics in Descartes' Natural Philosophy.", and Gaukroger and Schuster, "The Hydrostatic Paradox."

[49]Just as Descartes ignored the 'declaratory' Scholastic rules about the subordination of mixed mathematics, he ignored Stevin's strictures on the mutual irrelevance of natural philosophy to mixed and practical mathematics.

[50]The dynamic of research and concept formation unleashed here played out well beyond the optical work of the 1620s and extended directly to the important and little understood details of his vortex celestial mechanics in *Le Monde*, see Schuster, "'Waterworld': Descartes' Vortical Celestial Mechanics".

[51]Material in this case derives from Schuster, "Descartes *Opticien*".

via the most important and fruitful physico-mathematical research Descartes ever attempted: his work in geometrical and physical optics in the 1620s. This involved his discovery of the law of refraction of light around 1627, followed immediately by his exploration of possible mechanical rationales or explanations for the law. The latter attempts in turn were intimately connected with the process by which he crystallized his concepts of dynamics directly out of a 'physico-mathematical' 'reading' of his geometrical optical results.

In 1626/7 Descartes, collaborating with Claude Mydorge, constructed a law of refraction, by working within traditional geometrical optics in the limited mixed mathematical sense and without any corpuscular-mechanical theorizing. Descartes and Mydorge, like Harriot earlier, used the traditional image locating rule in order to map the image locations of point sources taken on the lower circumference of a half-submerged disk refractometer. (Fig. 3.6) Even using Witelo's fudged data, one gets a smaller semi-circle as the locus of image points, yielding a law of cosecants. In order to create a refraction predictor, Mydorge flipped the inner semi-circle up above the interface as the locus of point sources for the incident light. (Fig. 3.7).

On this representation of the new law Descartes then worked his favored style of physico-mathematical magic: Looking for a physics of light to explain the law, he transcribed into dynamical terms the geometrical parameters embodied in this diagram. The resulting dynamical principles concerning the mechanical nature of light were: (1) that the parallel component of the force of a light ray is unaffected by refraction, whilst (2) the quantity of the force of the ray is increased

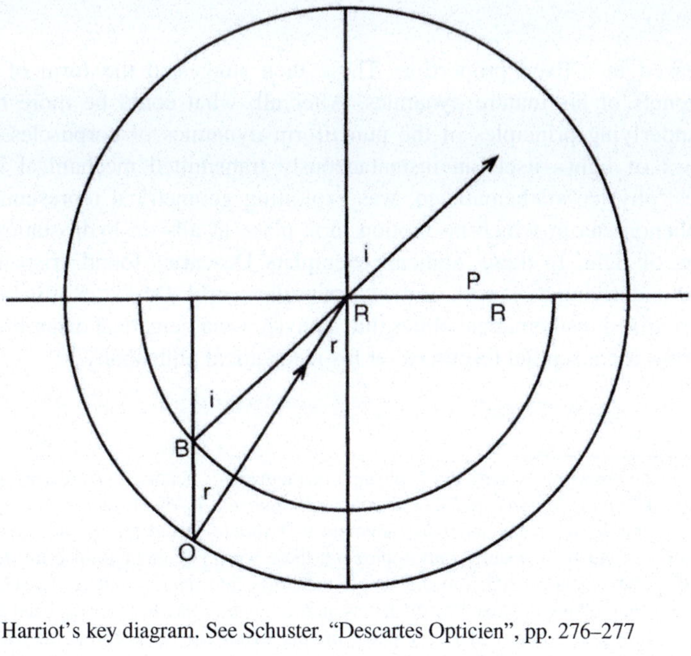

Fig. 3.6 Harriot's key diagram. See Schuster, "Descartes Opticien", pp. 276–277

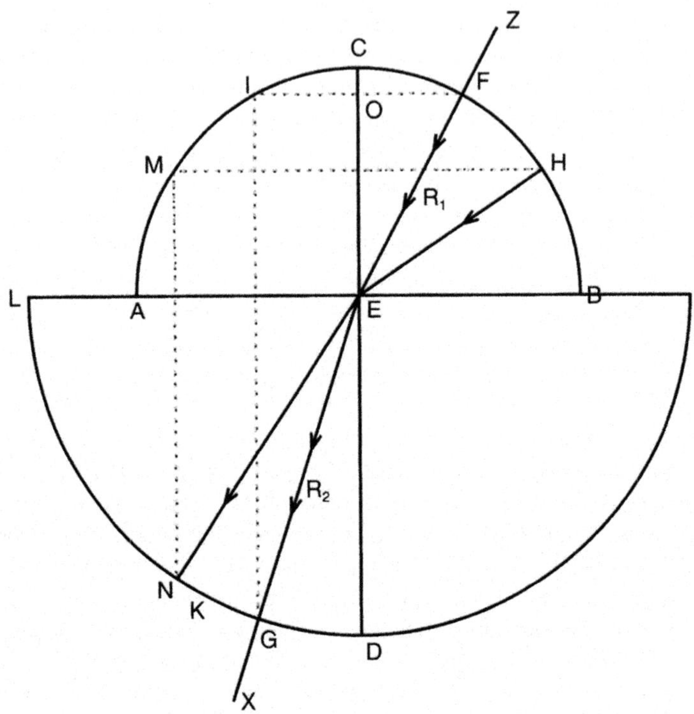

Fig. 3.7 Mydorge's refraction prediction device. Schuster, "Descartes Opticien", pp. 272–274

or decreased in a fixed proportion. These then suggested the form of the two central tenets of his mature dynamics. After all, what could be more revealing of the underlying principles of the punctiform dynamics of corpuscles than the basic laws of light—itself an instantaneously transmitted mechanical impulse? Descartes, physico-mathematician, was exploiting geometrical representations of telling phenomena in which no motion took place at all—in hydrostatics, and in refraction of light. In these 'statical' exemplars Descartes found crisp messages about the underlying dynamics of the corpuscular world. Descartes was bidding to transform mixed mathematical optics into a physico-mathematical discipline, and to extract from it conceptual resources for his mechanical philosophy.[52]

[52]There were competing varieties of physico-mathematics: Schuster, *Descartes–Agonistes: Physico–Mathematics, Method and Corpuscular–Mechanism*, 56–59. In addition to Descartes' program and the Jesuit mathematicians' attempts to promote mixed mathematics as 'separate but more or less equal' to natural philosophizing; there were [1] attempts to bring mechanics, particularly a dynamical approach to the simple machines into natural philosophy; [2] Kepler's profound neo-Platonizing of mixed mathematics and redirecting the thus physicalized disciplines back into natural philosophy; [3] Beeckman's linking of an emergent corpuscular mechanism to dynamical interpretations of the simple machines [Gaukroger and Schuster, "The Hydrostatic

According to the young Descartes' physico-mathematical strategy, any rigorous result in the mixed mathematical sciences bespeaks the discovery of a deep physical truth which can be reduced to corpuscular-mechanical terms.[53] Results in the mixed mathematical sciences are thus reduced and explained, and by extension, the further realms of practice are subsumed and controlled. These optical researches marked the high point of his work as a physico-mathematician transforming the 'old' mixed mathematical sciences and co-opting the results into a mechanistic natural philosophy. His optical results both confirmed his 1619 agenda of developing a corpuscular ontology and a causal discourse, or dynamics, involving concepts of force and directional 'determinations', and they shaped his conception of light as an instantaneously transmitted mechanical tendency to motion, as well as the precise principles of his dynamics.[54]

These examples are significant in the trajectory of Descartes, but in the larger process of the Scientific Revolution, they are small events. However, they do show how our reformed notions of 'source', 'target' and 'mode of causation' can illuminate specific episodes within the general theme of "what did practical and mixed mathematics have to do with the Scientific Revolution?" Let us take this approach further into the explanatory challenge of this volume, seeking bigger game through two more case studies.

3.5 Case Study 3: Sorting Out the 'Causal Mode' of Sixteenth-Century Mechanics

A common story of 'source, mode of causation, and target' stars sixteenth-century mechanics, a dynamic province of mixed mathematics: Sixteenth-century mechanics provided core concepts, metaphors, or values for the mechanical philosophy, or, for 'the new science' more generally. In this regard scholarly attention has recently focused on one strand of sixteenth-century mechanics, the pseudo-Aristotelian*Mechanical Questions*. For example, Helen Hattab, a leading scholar of Renaissance mechanics and philosophy, articulating the work of Rose, Drake and Laird, has documented how some sixteenth-century commentators on the *Mechanical Questions* tried to collapse the distinction between physical explanations of natural phenomena and geometrical explanations of machines, thus inviting mathematics into discourse concerning physical causation.[55] This is the

Paradox", 555–7]; finally [4] Galileo's rather more piecemeal physico-mathematical excursions, including his construction of a *sui generis* new kinematical science of motion.

[53] Gaukroger and Schuster, "The Hydrostatic Paradox", 568–70; and Schuster, "Descartes, *Opticien*", 279–85, 290–95.

[54] The optical work was indeed the technical high point of his physico-mathematical agenda, but the trajectory into natural philosophical systematics carried Descartes even further, to the 'completion' of this trajectory in the formulation of the vortex mechanics in *Le Monde* as I have argued in Schuster "'Waterworld': Descartes Vortical Celestial Mechanics" and Schuster, *Descartes–Agonistes: Physico–Mathematics, Method and Corpuscular–Mechanism*, chapters 4 and 10.

[55] Hattab, "From Mechanics to Mechanism".

type of process that should interest us regarding "practical mathematics and the Scientific Revolution". Hattab speculates that these border crossings shaped Descartes' approach to mechanics and mechanical philosophy, "influencing" him to absorb mechanics into physics, and apply mechanics to corpuscles, adding that Descartes derived general mechanical principles from analyzing one of the canonical "mechanical problems" in the text, the sling.[56]

Let us submit Hattab's speculative story to an exercise in rectification of explanation. I do this not because of any shortcomings in Hattab's scholarship, which is superb, but rather because her speculation resembles other "source, mode, and target" stories common in this area, and we are now well placed to unpack it. First of all, we should recognize that sixteenth-century mechanics *per se* exerted no "influence" or "imprinting" upon Descartes. Second, as we have seen, Descartes' dynamics was forged in his physico-mathematical hydrostatics and optics. It did not arise via the *Mechanical Questions*, nor was the sling the source of Descartes' dynamical concepts; rather, it illustrated them. Third, in support of Hattab, we can say sixteenth-century mechanics was indeed just about the first site where attempts were made, on the level of both declaratory policy and technical practice, to move a mixed mathematical field, closely linked to practical mathematics, into direct contact with natural philosophical issues of matter and cause. Fourth, Descartes certainly did some appropriating and translating. He picked up and re-emitted the legitimatory rhetoric of sixteenth-century mechanics to package detailed, technical work, not deducible from that legitimatory discourse. Those technical resources came from Stevin and geometrical optics. The young Galileo, by contrast, had dipped into Archimedes as well as the *Mechanical Questions* tradition for both sorts of resources.[57]

In sum, as signaled earlier, although technical developments have as their necessary causes technical resources and skills, rhetorical transactions must also be woven into rectified developmental stories, because they are crucial to actors' self-understandings, accountings of their own and others' claims and in general the enrolment of audiences. In explanations of technical and legitimatory developments, influence and imprinting should be avoided in favor of some version of a cultural process model, keyed to suitable conceptualizations of the traditions and fields in play.

[56] *Ibid.*, pp. 122, 126–7.

[57] None of this is intended to underplay the role of sixteenth-century developments in mechanics in the eventual crystallization of the classical mechanics of Galileo and Newton. Recall our observation that the expression 'mathematization of 'Science" should be construed as 'physicalization of the mixed mathematical sciences'. The attempt to 'upgrade' mechanics to natural philosophical status is a key strand in that long process. The construction of classical mechanics involved various strands, in many of which there were 'physico-mathematical' plays by mathematically oriented natural philosophers into the realm of mixed mathematics, for the purpose of physicalizing them and enhancing their relevance to natural philosophical issues of matter and cause.

3.6 Case Study 4: Hobnobbing with Practitioners and Machines

During the years he was writing *Le Monde* and living in the United Provinces, Descartes tried to design a machine to grind parabolic lenses. It differed slightly from the machine described later in the *Dioptrique*. He attempted to persuade the artisan Jean Ferrier to come join him in the project, and a technical correspondence ensued.[58] What do these transactions say about practical mathematics/natural philosophy relations, and about Descartes' strategies regarding the two fields?

First, Descartes was, in his fashion, making a play inside the field of practical mathematics. He did indeed want to make and 'show' lenses that would embody his law of refraction, and control an improved telescope. Such behavior is indistinguishable from that of an elite mathematical practitioner. But, second, he was maneuvering within the culture of natural philosophy: The lens grinding machine was also a physical/mechanical instantiation of the law of refraction[59] — not just a bid for fame and profit. Indeed, it was a natural philosophical signifier, denoting a concrete and specially valued achievement. His lens machine, guided by natural philosophical principles, surpassed anything that could have been produced by crafty trial and error, and as Ramus and Bacon would have acknowledged, it was both illustrative of the truth and maximally useful.[60]

It should also go without saying that Descartes had not killed off natural philosophy in the interest of modern experimental "method" or "science". Ferrier, who had worked with Descartes and Mydorge in the 1620s, came again into potential play regarding the new machine in the early 1630s after the construction of the central concepts of Descartes' dynamics, and as *Le Monde* was being written. Wanting to hobnob with Ferrier was not driving Descartes' natural philosophical

[58]Schuster, "Descartes–Agonistes: Physico–Mathematics, Method and Corpuscular–Mechanism", 401–403; William Shea, *The Magic of Numbers and Motion. The Scientific Career of René Descartes* (Canton, Ma, 1991), 191–201. These transactions are not to be confused with the work Ferrier actually undertook with Descartes and Mydorge regarding refraction earlier in the 1620s. (Schuster, "Descartes, Opticien", p. 272; Shea, *The Magic of Numbers and Motion*, 150–2.)

[59]I am pleased to point out that the late Michael S. Mahoney first made this point to me many years ago in informal discussion.

[60]Rossi, *Philosophy, Technology and Arts in Early Modern Europe*, masterfully established this general perspective. My points here relate to the putative signification of the lens-grinding machine as such. Neil Ribe interestingly widens this perspective, by demonstrating that for Descartes the ultimate aim of optical knowledge, practically embodied in telescopes and microscopes, is the improvement of (inherently limited) unaided human vision, in aid of the improvement of genuine knowledge to the purpose of generalized human mastery of nature. Ribe reminds us that at the conclusion of the *Diotprique* Descartes called for a new kind of artisan, from amongst the ranks of the "more curious and skilful persons of our age..." Niel Ribe, "Cartesian Optics and the Mastery of Nature," *Isis* 88 (1997): 42–61, at p. 61. See also the important insights of Jean-François Gauvin which extend even further Ribe's insights into the likely aims of the *Dioptrique*. Jean-Francois Gauvin, "Artisans, Machines and Descartes's *Organon*," *History of Science* 64 (2006): 187–216, at pp. 198–201.

agenda, inscriptions or strategies at all. Inside natural philosophy the instrument was *for* natural philosophical agendas and actions. Descartes had not stopped being a natural philosopher and become a new kind of 'scientist' because he played with instruments and instrument makers. He played with instruments and instrument makers because this fitted his evolving agenda as a natural philosophical contender. In short Descartes wished to position himself as the leading philosopher of nature, by means of strategically crucial articulation with, and appropriation of, the rhetoric, as well as the findings, practices and artifacts of practical mathematics.[61] The general historiographical lesson here follows from our cultural process model: Suppose we ask, 'What were instruments and their makers *for* inside natural philosophy?' The answer is, they were *for* natural philosophizing, *for* natural philosophers' agendas and actions. If we forget that, essence and origin stories will loom up, clouding our historiographical imaginations.

Finally, there is another lesson here for handling claims about the "influence" of the rhetoric and values of mathematical practitioners, because we are dealing with concrete "cultural process" transactions in a specific case. We can temper any claim that Descartes was "influenced" by mathematical practitioners by seeing how the values and rhetoric he appropriated geared into his process of work on a specific natural philosophical project. Hence we can calibrate what can and cannot be attributed to such a vague "influence" as the rhetoric, values or ideology of the mathematical practitioners. So, first of all, it is entirely possible Descartes appropriated practitioners' rhetoric and that this was used to express to others—and to himself—what he was doing and why. But, Descartes was doing more than practicing rhetoric. He was also "doing" optics, and "doing" natural philosophy in specific technical ways. Those "doings" are not deducible from the practitioners' rhetoric, caused by it, or influenced by it. Descartes appropriated the rhetoric to wrap his results in cultural understandings, attractive and persuasive to his audience, and importantly to himself as well, for thematizing his own roles and strategies. After all, we have seen how important to him had been his personal twist on the contemporary identity category of *physico-mathematicus*.

[61] See also Gauvin, "Artisans, Machines and Descartes's *Organon*" for very significant findings about the larger dimensions of Descartes' encounters with the practical arts and artisans, and their shifts over time, from his earliest thoughts on method and *mathesis universalis* to the *Discours de la Méthode* of 1637. Gauvin's inquiry goes well beyond our concerns here, which are limited to what practical and mixed mathematics had to do with specific technical and rhetorical developments in Descartes' work in natural philosophy and its subordinate disciplines. Gauvin demonstrates the import of Descartes' reflections upon—and idealizations of—the work and *habituses* of practical artisans for his formulation of the grandest ambitions of his method discourse, and for the widest socio-cultural implications of the *Discours* in the context of the rise of the French absolutist state. He also implies that Descartes was signaling that artisans' practices should be guided by his method. I stress, however, that the strictest post-Kuhnian skepticism should be maintained as to the actual technical efficacy of any of Descartes' general statements about universal method. On my previous work on this topic, see above, Note 36.

3.7 Conclusion: Opportunities and Pitfalls

When thinking about 'practical mathematics and the Scientific Revolution', we encounter a proliferation of uncritical stories of multiple sources and targets, linked by unsatisfactory causal categories, such as influence or imprinting. The answer to 'multiple sources for each given target' and 'multiple targets for each given source' is not imposition of one story, or retreat to local studies only. We can rectify the terms of explanation by modeling both natural philosophy and practical mathematics as contested fields in process over time. In this way the strengths of a reformed externalism and the new historiography of practical mathematics can be realized, and their pitfalls avoided. Events within the field of natural philosophizing did not involve members being forced or shaped from the outside by practical mathematics. Rather, players within natural philosophy appropriated, translated and utilized resources from without, with the resulting complex pattern of claims and outcomes—intended and unintended—being played out in the field of natural philosophy over time.

Finally, the approach taken in this paper may allow us to resolve the problem vulgarly expressed as 'how did science become mathematical'. The issue was not the 'mathematization of science' but rather the 'physicalization of the traditional mixed mathematical sciences' by radical natural philosophical challengers to neo–Scholastic hegemony, challengers who were, amongst other things, willing to appropriate and translate rhetorical and technical resources from the tradition of practical mathematics. We examined Descartes' strategies and gambits in this regard; but, Descartes was only one player in a competitive early to mid-seventeenth-century natural philosophical environment, where appropriation and natural philosophical deployment of mixed and practical mathematics—under the category 'physico-mathematics'—was cutting-edge practice for some contenders. To conclude, therefore, 'the story of practical (and mixed) mathematics and the Scientific Revolution' is really the sum of largely yet to be written critical narratives of these activities.

Part II
What Did Practical Mathematics Look Like?

Chapter 4
Mathematics for Sale: Mathematical Practitioners, Instrument Makers, and Communities of Scholars in Sixteenth-Century London

Lesley B. Cormack

Abstract "Mathematics for Sale" investigates the location of mathematics within London, arguing that the place and community for practical mathematics provided the foundation for the coffee house culture of the seventeenth century. This paper examines mathematical lectures and especially instrument-makers both inside and outside the City walls. In particular, Thomas Hood's mathematical lectures started a trend of such events, followed by a proposal by Edward Wright and John Tapp to fund a lecture in navigation. Gresham College played a small part in housing mathematics instruction for the larger community, but it was in the instrument shops of instrument-makers like Elias Allen and Emery Molyneux that like-minded men gathered to discuss practical mathematics and instruments. Cormack identifies 85 different instrument-makers and mathematical practitioners with shops and rooms in London between 1550 and 1630. She thus discovers a vibrant practical mathematical community, whose members were gentry, scholars, merchants, instrument-makers, and navigators. This rich mathematical exchange laid the groundwork for the natural philosophical sociability of the seventeenth century. These early mathematically-minded men and their ideas, however, did not change natural philosophy in a direct way.

L.B. Cormack (✉)
Department of History and Classics, University of Alberta, Edmonton, AB, Canada
e-mail: lcormack@ualberta.ca

L.B. Cormack et al. (eds.), *Mathematical Practitioners and the Transformation of Natural Knowledge in Early Modern Europe*, Studies in History and Philosophy of Science 45, DOI 10.1007/978-3-319-49430-2_4

4.1 Introduction

Robert Hooke's diary gives us a fascinating glimpse into the London natural philosophical scene in the 1660s and 1670s.[1] Hooke was constantly meeting friends, fellow artisans and natural philosophers, in the coffee shops and roadways of Restoration London. There was clearly a vibrant informal natural philosophical community – in the coffee houses, trades' and builders' shops, architects' offices, and the Royal Society rooms at Gresham College. Historians such as Larry Stewart have pointed to this informal culture as the turning point towards an interest in and acceptance of natural philosophy, and as indicative of its growing importance in the making of modern public space.[2] They have argued that this 'coffee house culture' changed the way natural philosophy was studied.

This sort of intellectual exchange had been developing at least 60 years earlier, however, before the advent of the coffee house, before the Royal Society's foundation, even before the opening of Gresham College. Scholars, craftsmen, merchants, lawyers, and gentlemen were meeting in London by the 1580s and 1590s. Men who had gone to university and found others with a like-minded interest in mathematics and the natural world met again in London, sometimes at the Inns of Court, occasionally through the mathematical lectures springing up around the city, and often through a shared interest in applied mathematics at the instrument makers' shops and the docks. There, important intellectual and material exchanges took place, which emphasized utility and precision. These exchanges were both international in scope, and national in ideology.

The relationship between the scholar and craftsman was a personal one, based on geographical proximity, shared interests and the exchange of expertise, ideas and goods for sale. By examining the place of mathematics practice in London in the 1580s to 1610s, we discover a rich tapestry of mathematical practitioners and mathematical instrument makers who made London their home in this period, including such entrepreneurial instrument makers as Elias Allen and Emery Molyneux. Equally, we discover the enthusiasm of partakers of this mathematical culture, such as Gabriel Harvey, and the role played by this interaction in the promotion of mathematical thinking for the wider natural philosophical enterprise. The trade in mathematical knowledge and instruments in late sixteenth-century and early seventeenth-century London encouraged an interest in natural knowledge, in

[1]Robert Hooke, *The Diary of Robert Hooke*, eds. Henry W. Robinson and Walter Adams (London, 1935). For a discussion of Hooke's coffeehouse practice, see Lisa Jardine, *The Curious Life of Robert Hooke. The Man who measured London* (New York: Harper Collins, 2004).

[2]Larry Stewart, The rise of public science: rhetoric, technology, and natural philosophy in Newtonian Britain (Cambridge: Cambridge University Press, 1992); Trevor Levere and G.L'E. Turner, *Discussing Chemistry and Steam: The Minutes of a Coffeehouse Philosophical Society 1780–1787* (Oxford: Oxford University Press, 2002). See also Brian Cowan, *The Social Life of Coffee: The Emergence of the English Coffeehouse* (New Haven: Yale University Press, 2005).

the utility of such knowledge, and the creation of a varied and vibrant discursive community that would encourage the furthering of natural investigation in the years to come.

4.2 Mathematical Lectures

London was a busy metropolis in the last decades of the sixteenth century, both for the numerous and hard-working merchants and for those more interested in mathematical and natural philosophical pursuits.[3] The Inns of Court, Parliament, and the Royal Court all provided reasons for many young men and women to find their way to the city. Combined with a growing interest in trade, investment, and exploration, London was an increasingly attractive destination for young men from the country, fresh from university or their estates, eager to make their way in the world and to find communities of like-minded individuals. These new inhabitants of London, combined with skilled émigrés fleeing the religious troubles of the continent, ensured that there was both the expertise in mathematics and a ready market for this expertise. In the second half of the sixteenth century, a number of university-trained or self-taught men set themselves up as mathematics teachers and practitioners. These men, who we might call mathematical practitioners, sold their expertise as teachers through publishing textbooks, making instruments, and offering individual and small group tutoring. In the process, they argued for the necessity of practical knowledge of measurement, surveying, and mapping, among others, rather than for a more philosophical and all-encompassing knowledge of the earth.

Mathematical practitioners were a relatively new category of scientifically-inclined men, who first made their appearance in early modern Europe.[4] Mathematics was a separate area of investigation from natural philosophy and those interested in mathematical issues had usually tied such studies to practical applications, such as artillery, fortification, navigation, and surveying.[5] These mathematical practitioners became more important in the early modern period and provided a necessary

[3]Deborah Harkness, *The Jewel House. Elizabethan London and the Scientific Revolution* (New Haven: Yale University Press, 2007). For a more general discussion of early modern London, see Steve Rappaport, *Worlds within Worlds: Structures of Life in Sixteenth Century London* (Cambridge: Cambridge University Press, 1989) and Ian Archer, *The Pursuit of Stability: Social Relations in Elizabethan England* (Cambridge: Cambridge University Press, 1991).

[4]With some modification, I take here EG.R. Taylor's important classification of the more practical men *The mathematical practitioners of Tudor and Stuart England* (Cambridge: Cambridge University Press, 1954). For modern treatment of these crucial figures, see James A. Bennett, "The Mechanics' Philosophy and the Mechanical Philosophy." *History of Science* 24 (1986), 1–28 and Stephen Johnston, *Making Mathematical Practice: Gentlemen, Practitioners and Artisans in Elizabethan England.* (Cambridge: Ph.D dissertation, 1994).

[5]Mario Biagioli, "The Social Status of Italian Mathematicians, 1450–1600." *History of Science* 27 (1989), 41–95.

ingredient in the transformation of nature studies to include measurement, experiment, and utility.[6] Their growing importance was a result of changing economic structures, developing technologies, and new politicized intellectual spaces such as courts. Most were university-trained, showing that the separation of academic and entrepreneurial teaching was one of venue and emphasis, rather than background. Mathematical practitioners claimed the utility of their knowledge, a rhetorical move that encouraged those seeking such information to regard it as useful.[7] It is impossible to know the complete audience for such works, but English mathematical practitioners seem to have aimed their books and lectures at an audience of London gentry, merchants, and occasionally artisans.[8] It is probably this choice of audience that most influenced their emphasis on utility, since London gentry and merchants were looking for practicality and means to improve themselves and their businesses.[9]

Mathematical practitioners professed their expertise in a variety of areas, especially such mathematical applications as navigation, surveying, ballistics, and fortification. For example, Galileo's early works on projectile motion and his innovative work with the telescope were successful attempts to gain patronage in the mathematical realm.[10] Descartes advertised his abilities to teach mathematics and physics. Simon Stevin claimed the status of a mathematical practitioner, including an expertise in navigation and surveying.[11] William Gilbert argued that his larger philosophical arguments about the magnetic composition of the earth had practical applications for navigation.

[6]James A. Bennett, "The Challenge of Practical Mathematics." In *Science, Culture and Popular Belief in Renaissance Europe*, edited by S. Pumfrey, P. L. Rossi and M. Slawinski (Manchester: Manchester University Press, 1991), 176–190. Thomas Kuhn, "Mathematical versus Experimental Traditions in the Development of Physical Science," in *The Essential Tension: Selected Studies in Scientific Tradition and Change* (Chicago: University of Chicago Press, 1977), 31–65 provides an early attempt to claim a different history for mathematics and natural philosophy.

[7]Katherine Neal, "The Rhetoric of Utility: Avoiding Occult Associations for Mathematics through Profitability and Pleasure." *History of Science*, 37 (1999), 151–178 discusses some attempts to make mathematics appear useful.

[8]Thomas Hood's lecture, *A Copie of the Speache made by the Mathematicall Lecturer, unto the Worshipfull Companye present ... in Gracious Street: the 4 of November 1588* (n.d. 1588) is a good example. See Harkness for a discussion of the complex interactions among London merchants, artisans and scholars.

[9]As we see in W.R. Laird, Chap. 7 of this volume, mathematicians could deny practicality when the imagined audience was courtly.

[10]Of course, once Galileo successfully gained a patronage position, particularly with the Florentine Medici court, he left his mathematical practitioner roots behind and became a much higher status natural philosopher. Mario Biagioli, *Galileo, Courtier: The Practice of Science in the Culture of Absolutism.* (Chicago: University of Chicago Press, 1993).

[11]Descartes was, of course, Jesuit-trained. Peter Dear, *Discipline and Experience: The mathematical way in the Scientific Revolution* (Chicago: University of Chicago Press, 1995), 34.

In England, an early example of a mathematician using his expertise to improve the mathematical underpinning of these useful arts was Robert Recorde, employed by the Muscovy Company to give lectures and write a textbook in elementary mathematics in the 1550s.[12] Recorde's early foray was to be repeated, especially in London, by mathematical practitioners, many of whom, such as Thomas Hood and Edward Wright, demonstrated an interest in mapping and navigation explicitly.

These mathematical practitioners offered lectures, individual tutelage, and the instruments to explicate the mathematical structure of the world. Sometimes this was done on a completely entrepreneurial model, that is, where the practitioner set up his shingle and attracted clients through publishing and publicity. At other times, mathematics lectures were founded and supported by a small group of interested men, such as was the case with Thomas Hood.

4.3 Thomas Hood as the First London Mathematical Lecturer

Thomas Hood (1556?–1620) was the first mathematics lecturer paid by the city of London and thus fits the patronage model of mathematics lecturers. However, he also published and encouraged private pupils, and therefore is equally an entrepreneurial mathematics teacher. Hood had attended Trinity College, Cambridge, where he had received his B.A. in 1578 and his M.A. in 1581.[13] In 1588, Hood petitioned William Cecil, Lord Burghley, to support a mathematics lectureship in London, to educate the "Capitanes of the trained bandes in the Citie of London."[14] This was a complex proposition, since the Aldermen and Lord Mayor of London would be the ones paying the bills, but the Privy Council had to give its approval in order to allow the lectures to proceed.

[12]Lesley B. Cormack, "*The Grounde of Artes*: Robert Recorde and the Role of the Muscovy Company in an English Mathematical Renaissance", *Proceedings of the Canadian Society for the History and Philosophy of Mathematics*, Vol. 16, 2003: 132–138. Stephen Johnston, "Recorde, Robert (*c.*1512–1558)," in *Oxford Dictionary of National Biography*, ed. H. C. G. Matthew and Brian Harrison (Oxford: OUP, 2004); online ed., ed. Lawrence Goldman, January 2008, http://www.oxforddnb.com/view/article/23241 (accessed June 19, 2009).

[13]Biographical material on Thomas Hood can be found in Taylor, *Mathematical Practitioners*, 40–41; David W. Waters, *The Art of Navigation in England in Elizabethan and Early Stuart Time* (New Haven: Yale University Press, 1958), 186–189; H. K. Higton, "Hood, Thomas (*bap.* 1556, *d.* 1620)," in *Oxford Dictionary of National Biography*, ed. H. C. G. Matthew and Brian Harrison (Oxford: OUP, 2004), http://www.oxforddnb.com/view/article/13680 (accessed June 19, 2009).

[14]BL Lansdowne 101, f. 56.

Hood received the following positive response from the Privy Council:

> The readinge of the Mathematicall Science and other necessarie matters for warlike service bothe by sea and lande, as allso the above saide traninge shalbe continued for the space of 2 yeares frome Michaelmas next to come and so muche longer as the L. Maior and the Citie will give the same alowance or more then at this present is graunted.[15]

Hood's lectureship therefore went forward, held in the home of Sir Thomas Smith, merchant and later Governor of the East India Company. The makeup of the audience is now unknown, although from the tone of his introductory remarks, published under the title of *A Copie of the Speache made by the Mathematicall Lecturer, unto the Worshipfull Companye present ....in Gracious Street: the 4 of November 1588*, Hood seemed to be talking to his mathematical colleagues and mercantile patrons, rather than to the mariners he insisted needed training.[16] The contents of Hood's lectures are also unknown, but the treatises bound with the British Library copy indicate that he stressed navigational techniques, instruments, astronomy, and geometry.[17]

By 1590, Hood had been giving these mathematics lectures for almost two years, as he reported in his 1590 translation of Ramus.

> ... so that the time limited unto me at the first is all most expired ... In this time I have binne diligent to profite, not onlie those yong Gentlemen, whom comonlie we call the captaines of this citie, for whose instruction the Lecture was first under taken, but allso all other whome it pleased to resorte unto the same.[18]

Hood identified himself on the title-pages of all his books until 1596 as "mathematical lecturer to the city of London", sometimes advising interested readers to come to his house in Abchurch Lane for further instruction, or to buy his instruments.[19] His books explain the use of mathematical instruments such as globes, the cross-staffe, and the sector, suggesting that his lectures and personal instruction would have emphasized this sort of instrumental mathematical knowledge and understanding. While some historians have questioned what happened

---

[15]Ibid, f. 58.

[16]Hood (n.d. (1588)), sig. A2a ff.

[17]Thomas Hood, *The use of the two Mathematicall Instruments, the crosse Staff, ... And the Iacobs Staffe* (London, 1596); and *The Making and Use of the Geometricall Instrument, called a Sector* (London, 1598). BL 529.g.6.

[18]Petrus Ramus, *The Elementes of Geometrie.. translated by Thos Hood, Mathematicall Lecturer in the Citie of London* (London, 1590), sig. 2a.

[19]Thomas Hood lists himself as a mathematical lecturer on the frontispiece of the following books: *A Copie of the speache: made by the Mathematicall Lecturer unto the Worshipfull Companye present in Gracious Street the 4 of November 1588* (London, n.d.); *The Use of the Celestial Globe in Plano, set foorth in two Hemispheres* (London, 1590); *The Elementes of Geometrie. Written in Latin by that excellent scholler, P. Ramus, Professor of the Mathematicall Sciences in the Universities of Paris. And faithfully translated by Tho. Hood, Mathematicall Lecturer in the Citie of London* (London, 1590); *The Use of Both the Globes, Celestiall, and Terrestriall, most plainely delivered in forme of a Dialogue* (London, 1592); and *The use of the two Mathematicall Instruments, the crosse Staff, ... And the Iacobs Staff* (London, 1596).

at Hood's lectures (or if indeed they did happen), this larger evidence indicates both that there were such lectures, and that a number of leaders of the community, as well as mathematical practitioners like Hood, thought they were important in creating mathematical literacy and conversation in the City of London.[20] This was the beginning of a recognition of the power of mathematics for understanding the answers to practical problems and with it, a sense that mathematical answers were as legitimate as philosophical ones.

4.4 Gresham College

Any discussion of sites of mathematical instruction in London must include the role of Gresham College. While it would be a mistake to overemphasize the importance of this odd venture in adult education, Gresham provided a living to important mathematicians interested in the theory/practice exchange and was located in the vibrant commercial centre of the city. One of the peculiarities of Gresham is that it is a sixteenth-century conception, largely played out in the seventeenth century. Although Thomas Gresham technically founded the College in his will of 1579, the College could not be created until his widow's death in 1597.[21] It appears that Gresham College did not begin to deliver lectures until the seventeenth century, and therefore technically it was not part of the mathematical practice community of late Tudor London. As Ian Adamson put it, "It is a point sometimes overlooked by those whose eyes are on Gresham College but whose minds are on the Royal Society, that the College and its subject-range emanated from a man who was born 140 years before the Royal Society was founded."[22]

We know remarkably little about Gresham College – who attended lectures, or how many lectures were actually given, for example. Most of the surviving documentation deals with the acrimonious relationship between the professors and the governors through much of the College's history, rather than shedding any light on the actual teaching of the College. That said, there were a number of professors who appear to have given lectures on mathematics and navigation, especially Henry Briggs, Professor of Geometry 1596–1620 and Edmund Gunter,

---

[20]Further evidence of Hood's lectures is the fact that John Stow mentions them in his *Survey of London* (London, 1598), 57.

[21]Ian R. Adamson. "The Administration of Gresham College and its Fluctuating Fortunes as a Scientific Institution in the Seventeenth Century," *History of Education* 9 (1980):13–25. Earlier discussions of Gresham, linking it to the formation of the Royal Society and the new science, included Christopher Hill, *Intellectual Origins of the English Revolution* (Oxford: Clarendon Press, 1965) and Francis R. Johnson "Gresham College: Precursor of the Royal Society," *Journal of the History of Ideas* I: 413–38.

[22]Ian R. Adamson, "Foundation and early History of Gresham College, London, 1596–1704," (PhD. Dissertation, Cambridge University, 1976), 249.

Professor of Astronomy 1619–1626.[23] Briggs, for example, explained in 1616 that "I have publickly taught the meaning and use of this booke [i.e. Napier's Logarithms] at Gresham house."[24] Likewise, both Briggs and Gunter were named as accessible experts on practical mathematical issues, and Gunter's inventions of calculational devises were designed to aid those with less facility or knowledge of the underlying concepts.[25] Thus, although Gresham College was probably not the most important source for the interchange of mathematical ideas in London, the existence of Gresham did contribute to mathematical exchanges in the City and provided a livelihood and a focus on practical mathematics for several important mathematicians. Still, it is not at Gresham per se that we see the development of a strong mathematical culture likely to transform the investigation of the natural world; rather, mathematical conversations were taking place at lectures and in artisanal workshops.

4.5 Proposed Lecture in Navigation

A number of mathematically-minded men felt that the lectures in mathematics as performed by Hood and Briggs were not particularly useful for furthering the imperial aims of the English nation, in large part because they were too theoretical, and began to petition for more applied lectures, to be held at more convenient times. The petitioners saw the importance of these lectures as an opportunity to educate those without the benefit of either a university education or the leisure and means necessary to teach themselves. Those so educated would transform not natural philosophy, but rather the practice of navigation, surveying, fortification, and the like – the backbone of empire-making and entrepreneurial commerce.

Edward Wright, in the second edition of *Certaine Errors in Navigation* (1610), urged Henry, Prince of Wales: "and cease not there [with reforms of Trinity House], but intend also to found the long wished for Lecture of the Art of Navigation".[26] In 1613, John Tapp made this proposal even more explicit. In his "Dedication to Sir Thomas Smith", Tapp thanked Smith for the lecture that he already funded, in his

---

[23]Lesley B. Cormack, *Charting an Empire. Geography at the English Universities 1580–1620* (Chicago: University of Chicago Press, 1997), 204–6. For biographies, see Wolfgang Kaunzner, "Briggs, Henry (*bap.* 1561, *d.* 1631)," in *Oxford Dictionary of National Biography*, ed. H. C. G. Matthew and Brian Harrison (Oxford: OUP, 2004), http://www.oxforddnb.com/view/article/3407 (accessed June 19, 2009) and H. K. Higton, "Gunter, Edmund (1581–1626)," in *Oxford Dictionary of National Biography*, ed. H. C. G. Matthew and Brian Harrison (Oxford: OUP, 2004), http://www.oxforddnb.com/view/article/11751 (accessed June 19, 2009).

[24]Henry Briggs, preface to John Napier, *A Description of the Admirable Table of Logarithmes* (London, 1616), A6a.

[25]Cormack, *Charting*, 119, 205. Higton, "Gunter." Mordechai Feingold, *The Mathematician's Apprenticeship: Science, Universities and Society in England, 1560–1640* (Cambridge: Cambridge University Press, 1984), especially 33, 50–52, 69, and 138.

[26]Edward Wright, *Certaine Errors in Navigation* 2nd edition (London, 1610), f.*9b.

own home (which by this time would have been that given by Edward Wright, for the East India Company[27]), but said that these lectures have been to

> little profit as may be guessed, by the little Audience which doe commonly frequent them ... for the Arts there taught, I meane the Mathematiques, the practisers thereof are few, in respect of those that are practisers and professors of Navigation, which are generally all the better sort of Marriners and Sea-men, (and those practisers aforesayd) beejng for the most part either Gentlemen of the Countrey, or such in the Cittie, whose Law busines or other occasions in the tearme time, hinders them from those exellent exercises, which doubtles all of them doe zealously love and applaud.
>
> But were there a lecture of Navigation, a profession which a multitude of people make their onely living by. And to be read in such a place, where they shall not onely bee seene, knowne, and noted, (for wellspending their time,) by their owners, setters foorth and principall emploiers, but also their daily and frequent busines attracting and necessarily drawing them thither, there is no question to be made of a very sufficient Auditory and great benefit to be reaped thereby, as doubtles much good hath already been affected by the late readings, in lesse frequent & eminent places before time.[28]

In other words, Tapp was suggesting that the subject of the present lectures was too theoretical and therefore aimed at gentry and professional men. Given that this audience had other more pressing occupations, the numbers attending these lectures were low. But, as Tapp argues, this was the wrong emphasis and the wrong audience in any case. How much better to teach applied and practical mathematics, to the people who would use it every day – navigators, particularly. Further, how much more satisfactory to hold these lectures in a venue where people could be seen to be attending and therefore where one could use public shame or praise to entice attendance. Just as was the case with Hood's lectures, the elite imagined the good such lectures would do for their inferiors. The inferiors (and superiors) had other ideas, however. The navigation lecture seems not to have been founded. Perhaps most of these more practical men were getting the mathematical education they needed in more convenient locales for such mathematics – at the instrument makers' and sellers' workshops around London.

4.6 Instrument Makers

Those who gave and attended these mathematical lectures had some expectation that they would be able to buy, sell and use the instruments that were there discussed. It is no surprise, therefore, that just as these lectures were being presented to the London community, mathematical instrument makers were beginning to ply their trade in increasing numbers in late sixteenth-century London.[29] Until recently, few mathematical and philosophical instrument makers were known to have worked

[27] A. J. Apt, "Wright, Edward (*bap.* 1561, *d.* 1615)", in *Oxford Dictionary of National Biography*, ed. H. C. G. Matthew and Brian Harrison (Oxford: OUP, 2004), http://www.oxforddnb.com/view/article/30029 (accessed June 19, 2009).

[28] John Tapp, *The Path-Way to Knowledge* (London, 1613), sig A2b–A3b.

[29] Harkness, chapter 3, 97–141.

in London in the sixteenth century and those known were mostly foreign-born artisans plying their trade. Historians of instrument-making have argued that the huge expansion of English instrument-making occurred during the late seventeenth century, when English instruments became as good as any in Europe.[30] All accounts of the development of instrument-making in England point to Thomas Gemini, a goldsmith, as the first English instrument-maker in the 1550s, followed by Humfry Cole in the 1580s.[31] The line of engravers and instrument makers, beginning with Augustine Ryther and carrying through Charles Whitwell to the famous early seventeenth-century instrument-maker, Elias Allen, completes the typical roster of sixteenth- and early seventeenth-century instrument makers.[32] This is at best an incomplete picture, however. There were many more men involved in the making and disseminating of precision mathematical instruments in this earlier period.[33] I have discovered 26 instrument makers and sellers living in England between 1550 and 1600, and a further twelve mathematical practitioners (those who sold their expertise with these instruments). There were another thirty-seven such individuals in London between 1600 and 1630, for a total of 85 identified in this trade in an 80 year period.[34]

Part of the reason for the relative invisibility of these men is the lack of historical artifacts left behind. For some of these artisans, we have only a mention in a book, either printed in the book itself or, more often, a handwritten note by its reader,

---

[30] G.L.'E Turner, "The Instruments Makers of Elizabethan England," *Sartoniana* [Ghent], 8 (1995): 19–31; *Elizabethan instrument makers: the origins of the London trade in precision instrument making* (Oxford: Oxford University Press, 2000).

[31] Gloria Clifton, *Directory of British Scientific Instrument Makers 1550–1851* (London: Philip Wilson Publishers, 1995), introduction xi–xii, 111. Silke Ackerman, ed., Humphrey *Cole: Mint, Measurement, and Maps in Elizabethan England.* B.M. Occasional Paper, Number 126 (London, 1998).

[32] Elizabeth Baigent, "Ryther, Augustine (*d.* 1593)," in *Oxford Dictionary of National Biography*, ed. H. C. G. Matthew and Brian Harrison (Oxford: OUP, 2004), http://www.oxforddnb.com/view/article/24428 (accessed June 19, 2009); H. K. Higton, "Allen, Elias (*c.*1588–1653)," in *Oxford Dictionary of National Biography*, ed. H. C. G. Matthew and Brian Harrison (Oxford: OUP, 2004); online ed., ed. Lawrence Goldman, January 2008, http://www.oxforddnb.com/view/article/37108 (accessed June 19, 2009).

[33] James Bennett makes the distinction between mathematical instruments, that is, those that essentially measure, and philosophical instruments, those devised to test or explain underlying explanations of the world. "Knowing and Doing in the Sixteenth Century: What were Instruments for?" *British Journal for the History of Science* 36 (2003), 129–150. While there are some problems with this categorization (see Deborah Warner, "What is a Scientific Instrument, When did it Become One, and Why?" *British Journal for the History of Science* 23 (1990): 83–93), I would contend that all the instruments I am discussing are mathematical rather than philosophical.

[34] These instrument-makers were identified through a variety of sources. Some, like Thomas Hood, advertised their shop and expertise in their own books. Others were named in print by mathematical authors (for example, Charles Whitwell, is named in both Hood, *Sector*, title page, and William Barlow, *The Navigator's Supply Conteining many things of Principall importance* (London, 1597), title page). Edward Worsop names John Bull, John Reade, and James Lockeson, *A Discoverie of sundrie errours and faults* (London, 1582), A3b. Finally, manuscript citations supply further names, most famously of course the marginalia of Gabriel Harvey (see note 41).

suggesting the best places to buy instruments in brass or wood. About others, however, there is more information, recently pieced together through the painstaking work of historians such as Joyce Brown or Silke Ackerman.[35] Much more needs to be done, but the beginning of a picture of these instrument makers and mathematical practitioners in London is emerging.

The first place to look for artisans in the city of London is within its guilds. London guilds controlled almost all manufacturing and commerce within the city walls and in order to be a legitimate and prosperous merchant, entry into a guild was a necessary prerequisite.[36] However, there was no guild particularly focussed on instrument-making, largely because this was a new enterprise. How could new industries break into the restrictive and regulated world of London commerce? One choice was to operate outside the walls. The other was to use pre-existing guilds, and use their structure rather than their trade secrets. The latter seems to have been the strategy for most instrument makers. We thus find instrument makers in a number of London guilds, none of which would seem obvious for precision instrument manufacture and sale. Humfry Cole seems to have been one of the few instrument makers who was a goldsmith, for example, which probably had more to do with his work at the Mint than with his use of precious metals for the instruments.[37] Joyce Brown has discovered a dynasty of mathematical instrument makers connected to the Grocers guild. Specifically, Augustine Ryther was a Grocer and, by apprenticing Charles Whitwell, who in turn apprenticed Elias Allen, they created an impressive line of Grocer instrument makers.[38] This did not mean, however, that Grocers had a monopoly, since there were instrument makers who were freemen in the Stationers, Broderers', and Joiners as well.[39] Some, such as Emery Molyneux, worked outside the walls, where guild control was more attenuated. Just as mathematical lecturers and mathematical practitioners were entrepreneurial, so too were mathematical instruments makers, who worked within the system and without to find a foothold for this new industry.

Equally, mathematical instrument makers were located in strategic areas of London. There were three main centres for the workshops of mathematical instrument makers and mathematical practitioners in the city of London (see Fig. 4.1 for a map of the locations):

[35]Joyce Brown, *Mathematical Instrument-Makers in the Grocers Company 1688–1800* (London, 1979). Silke Ackermann, "Cole, Humfrey (*d.* 1591)," in *Oxford Dictionary of National Biography*, ed. H. C. G. Matthew and Brian Harrison (Oxford: OUP, 2004); online ed., ed. Lawrence Goldman, January 2008, http://www.oxforddnb.com/view/article/5853 (accessed June 19, 2009).

[36]Ian Anders Gadd, *Guilds, society & economy in London, 1450–1800* (London, 2002) and ed. *Guilds and association in Europe, 900–1900* (London, 2006).

[37]Ackermann, "Cole".

[38]Joyce Brown, *Mathematical Instrument-Makers in the Grocers Company 1688–1800* (London: Science Museum, 1979).

[39]*Ibid.*, 5.

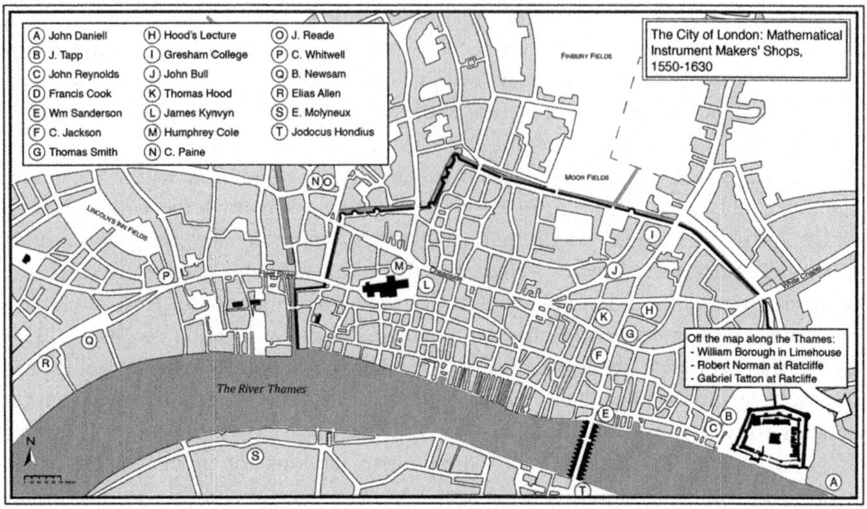

Fig. 4.1 Instrument makers and sellers in London, 1550–1630

1. Just west of London, especially north of the Inns of Court
2. In the City proper, around the Royal Exchange
3. East of the city, at the docks, especially at Limehouse and Ratcliff

Interestingly, there were no instrument makers in St. Paul's Churchyard, the location of many booksellers. Thus, we should not see instrument makers and book sellers as an interrelated group, even though they might well have shared similar clientele. Indeed, these three locations suggest that there were different types of clientele interested in mathematical instruments. The proximity to the Inns of Court supports the connection between gentlemen reading the law and an interest in exploration, exemplified by men such as William Crashawe, who was a member of the Virginia Company and a supporter of mapping and navigation.[40] The City site, close to the Royal Exchange and Gresham College, was in the heart of the commercial and financial district. Finally, those instrument makers at the docks might have been catering to a more practical clientele, due to set sail and needing the instruments that would assist them on their voyages. It is possible, although not known, that these latter shops might have specialized more in instruments in wood, rather than the more expensive and prestigious brass instruments, perhaps more available near the Inns.

[40]Fisher, R.M. "William Crashawe and the Middle Temple Globes 1605–15". *Geographical Journal* 140 (1974): 105–112.

4.7 Practical Mathematics and Its Audience

We know about these various instrument makers because mathematical practitioners and clients recommended them. Mathematical authors recommended in their books the best place to buy the instruments they described within their pages, and mathematical instrument makers were often authors themselves of manuals of the instruments they were selling. In fact, the distinction between those devising the instruments, or the understanding of the world implied by the instruments (the more 'scholarly' participants), and the maker of the instruments (the 'craftsman') was blurred. There was no 'scholar/craftsman' divide here, since the relationship among maker, explainer, and consumer of these instruments was constantly shifting. For example, Edmund Gunter (a scholar) named Elias Allen, J. Thompson, and N. Gos as instrument makers in *De sectore & radio. The description and vse of the sector in three bookes* (1623). Charles Whitwell is recommended by both Thomas Hood (a scholar and mathematical practitioner) and William Barlowe (a mathematical practitioner).[41]

A virtuoso like Gabriel Harvey could note (on the title page of his copy of John Blagrave, *The Mathematical Jewel*, himself a self-described "Gentleman and well willer to the Mathematickes"):

> Mr. Kynvin selleth this Instrument in brasse.

and

> His familiar Staff, newly published this 1590. This Instrument itself, made & solde by M. Kynvin, of London, neere Bowles. A fine workman, & mie kinde frend: first commended unto me bie M. Digges, & M. Blagrave himself. Meaner artificers much praised bie Cardan, Gauricus, & other, then He, & old Humfrie Cole, mie mathematical mechanicians. As M. Lucar newly commendes Jon Reynolds, Jon Redd, Christopher Painc, Londoners, for making Geometrical Tables, with their feet, frrames, rulers, compasses & squires. M. Blagrave also in his Familiar Staff, commendes Jon Read, for a verie artificial workman.[42]

In this latter note, we can see a community of interested men, sharing tips and knowledge. Diggs, Blagrave and Lucar were mathematical practitioners and renowned authors. The others named, including 'old Humfrie Cole', the patriarch of instrument makers, were London-based artisans. Harvey, a gentleman and virtuoso, could refer to some of these instrument makers as 'mie kinde frend', indicating the close connections that crossed any class boundary. The picture that emerges from this note is a close and interconnected community of practical mathematics. Interested gentry, mathematical practitioners, investors, and navigators interacted through their instruments. Arguably they gathered at the shops of many of these

---

[41] See note 33.

[42] Gabriel Harvey, marginal note signed by Harvey, with date 1585 in BL c.60.o.7 John Blagrave, *The Mathematical Jewel* (1585), title page. At bottom of page for first quotation, on right side for second. For a more complete discussion of Harvey's marginalia, see Virginia F. Stern, *Gabriel Harvey: His Life, Marginalia, and Library* (Oxford: Clarendon Press, 1979).

instrument makers, where they debated the merits of different instruments, looked at the latest 'universal instrument' that would measure all things and explain the astronomical movement of the heavens, and exchanged gossip about new explorations and star catalogues. In the process, these men began to develop an identity with a defined interest group, an identity that led to a boost in the self-mage of the instrument-makers and their acceptance by the interested gentry.[43]

A good illustration of such a meeting-place was the posh shop of Elias Allen, whose location out of the city on the Strand indicates his connections with the Court and Parliament at Westminster and the importance of the growing residential and cultural West End. Allen had first been apprenticed to Charles Whitwell of the Grocers' Company, and went on to become a freeman of that company, indicating his status within the London elite.[44] Allen became perhaps the best known mathematical instrument maker in London in the early seventeenth century. For example, he was recommended by Arthur Hopton in the pages of his *Speculum Topographicum* in 1611.[45] He made instruments for James I and Charles I; he was associated with Edmund Gunter and William Oughtred, professors of Gresham College and themselves eminent mathematicians. His workshop was used as "a general meeting place and also as a post office for letters between scholars". His shop was mentioned by William Oughtred, John Tuysden, and William Price as a place to meet in London.[46] This was elevated company for a 'mere' practitioner, and illustrates the significant transformation in status and self-confidence these precision instrument-makers were experiencing at this time. The continuing hub of activity at Allen's shop, combined with the high stature he gained during his lifetime, shows that mathematical instruments provided a focus for a wide range of men interested in the mathematical understanding of the world and in using mathematics to advance their economic and political plans. Here was the beginning of the so-called coffee house culture that brought together men who could and did change the interpretation of nature to include measurement and mathematics.

4.8 Molyneux's Shop

Emery Molyneux's shop provides a contrasting example of a smaller instrument-maker, a man working in relative isolation compared with Allen, and yet with a workshop where the politically and mathematically influential could still meet.

[43]Pamela Smith, *The Body of the Artisan. Art and Experience in the Scientific Revolution* (Chicago: University of Chicago Press, 2004), ch. 1, describes this growing sense of self-mastery and knowledge among the Flemish naturalistic painters of the fifteenth century.

[44]H. K. Higton, "Allen, Elias (*c.*1588–1653)," in *Oxford Dictionary of National Biography*, ed. H. C. G. Matthew and Brian Harrison (Oxford: OUP, 2004); online ed., ed. Lawrence Goldman, January 2008, http://www.oxforddnb.com/view/article/37108 (accessed July 18, 2009).

[45]Arthur Hopton, *Speculum Topographicum* (London, 1611).

[46]Hester Higton, "Portrait of an Instrument-Maker: Wenseslaus Hollar's Engraving of Elias Allen," *British Journal for the History of Science* 27(2) (2004), 155.

Molyneux's workshop was in Lambeth, south of the Thames and therefore off the beaten track. At this workshop he made instruments; we know for example that he made a Mariner's compass and a sector.[47] Richard Polter claimed that he was a good man with a lodestone, suggesting that navigators would visit his shop to 'set' their compasses before a long voyage.[48] Later in his career, Molyneux petitioned Lord Burghley for a patent for his new design of a canon.[49] And, of course, most famously, he made a pair of globes, the Molyneux Globes.

Emery Molyneux is a relatively unknown figure in instrument history, despite his well-known instruments. We know almost nothing of his family or how he came to be an instrument-maker. Nor do we know why he chose to set up shop in Lambeth, far from the hub of London instrument activity.[50] It may well be that he had some connections with the Archbishop of Canterbury, or some other means to gain access to that ward of London. What we do know is that by the 1580s, he was an established instrument-maker, with important connections with the elite voyagers of discovery. He knew important voyagers such as John Davis and Sir Walter Ralegh. He went on at least one extended voyage with Sir Francis Drake.[51] He had connections with William Sanderson, an important London merchant and overseas investor, who paid for the design and construction of the great Molyneux globes, and he worked on the globes with the university-trained Edward Wright and the Dutch engraver, Jodius Hondius.[52]

Molyneux's was an artisanal workshop in which he constructed his instruments, with visitors coming by to observe and discuss. Scholars such as Edward Wright and Richard Hakluyt had access to Molyneux's workshop, the former as a collaborator on the world map to be pasted to the face of the round globe and the latter as an interested observer and reporter. Hakluyt must have visited the shop several times, announcing in 1589 that globes should appear soon:

---

47 Helen M. Wallis, "The First English Globe: A Recent Discovery," *The Geographical Journal* 117 (1951), 277.

48 Richard Polter, *The Pathway to Perfect Sayling* (London, 1605), sigs. D1a–b.

49 BL Lansdowne 101, f. 69 (#17): "petition by Emery Molineux to Lord Burghley re discovery of a new demi-culverin, etc. 1596."

50 Susan Maxwell, "Emery Molyneux". *Oxford Dictionary of National Biography*, Vol. 38 (Oxford: Oxford University Press, 2004), 556–557.

51 John Davis claimed in *Hydrographical Discription* (London, 1595) that he suggested Molyneux' name to William Sanderson as an instrument-maker who could make globes desired by Sanderson. On these globes, Molyneux states: "I have been able to do this both in the first place from my own voyages and second from that successful expedition to the West Indies under the most illustrious Francis Drake in which expedition I have put together not only all the best delineations of others, but everything my own humble knowledge or experience has been able to furnish in the last 5 years to the perfecting of this work." Quoted in Wallis, "First English Globe", 279.

52 For further discussions of the making of this globe and its connection to larger cosmographical, political, and mathematical publics, see Lesley B. Cormack, "Glob(al) Visions: Globes and their Publics in Early Modern Europe," in *Making Publics in Early Modern Europe: People, Things, Forms of Knowledge*, Paul Yachnin and Bronwen Wilson, eds. (New York: Routledge University Press, 2009), 138–156.

> the coming out of a very Large and most exact terrestriall Globe, collected and reformed according to the newest, secretest, and latest discoveries, both Spanish, Portugall, and English, composed by M. Emmerie Mollineux of Lambeth, a rare Gentleman in his profession, being therein for divers yeeres, greatly supported by the purse and liberalities of the worshipfull merchant M. William Sanderson.[53]

In 1591, Petruccio Ubaldini, the Tuscan emissary in London, visited Molyneux's workshops and saw the globes being made, giving us an important first-hand account of the interactive space of the workshop for the exchange of scientific, technological and political information. Ubaldini reported back to the Tuscan court that he had observed Molyneux at work and talked to him about how the globes were constructed. Molyneux explained that he made the form with paste so that the globes would resist humidity. Ubaldini recounted that he was impressed by the size of the globes and the details they depicted, especially the paths of English voyages. He tried to purchase a pair from Molyneux, but was told that they were not for sale. Molyneux told him that he would give the first pair to the Queen, and that he then intended to take them himself to Europe where he would present them to several interested princes.[54] Here was an instrument-maker with a strong sense of confidence in his expertise and the quality of his instruments. As Ubaldini reported, "and that is why I could not find out the price."[55] Ubaldini evidently had geopolitical reasons for his interest, but was also a well-informed participant in the workshop community, a community made up of a growing number of men with interest in mathematics for its utility, its political application and perhaps its metaphysical import.

4.9 Conclusion

By 1610, there was a strong practical mathematical community living in London. Most significantly, the mathematical instrument trade, including the trade in maps and globes, had vastly expanded in the last years of the sixteenth century. A number of mathematical lectures had been sponsored, attended by a variety of audiences. Books and individual lessons explaining the use of mathematics and mathematical instruments had been produced, all leading to an increasing number of men trained in and sensitive to mathematical tools and explanations. A variety of men met in the instrument shops and at the mathematical lectures – gentlemen like Harvey, scholars like Edward Wright, merchants interested in investment like William Sanderson, and navigators intent on accurate information both to give and receive.

[53]Richard Hakluyt, in *Principal Navigations, Voyages, and Discoveries of the English Nation* (London, 1589), sig. *4b.

[54]The fact that two of the five extant Molyneux globes are now in Germany suggests that he may have done just that, although we have no evidence of this beyond their location.

[55]Anna Maria Crino and Helen Wallis, "New Researches on the Molyneux Globes". *Der Globusfreund* 35–7 (1987), 14.

This was the face of mathematical practice in late sixteenth- and early seventeenth-century England. Significant numbers of men had invested time, energy, and funds in the mathematization of practice and theory. However, this community of mathematically-minded men had not changed the philosophical interpretation of nature. Indeed, most of them were content to leave philosophizing to others and to concentrate instead on the practical improvements they could make with this mathematical knowledge and these instruments. What this huge effervescence of mathematics provided in this early period was not a transformation of natural philosophy, but rather a transformation of the sociological structure of nature studies. The interaction of these men from different social, economic, and vocational strata marked a major change in the scientific community. One entered this community through shared interests and through the sociability of the exchange of ideas and expertise. These social interactions of the late sixteenth century provided the basis for an expanded culture of debate and discourse in the seventeenth century. The introduction of the first coffee houses in London in the 1650s[56] did not create the sociability that led to the Royal Society and to the widespread acceptance of natural philosophical debate. Rather, these coffee houses were just another venue. Instrument shops, along with book shops, the Inns of Court, and the rooms in Gresham College (at least, until the College burnt down) continued in importance throughout the seventeenth century and well into the eighteenth.[57] The increasing wealth and disposable income of the virtuosi and other interested gentlemen and merchants ensured an expanding market for instruments, books, private instruction, and conversation in the years to come, seen as the 'coffee house culture' so important for Robert Hooke 60 years later.

[56]Cowan, *Social Life*, p, 25.

[57]James A. Bennett, "Wind-gun, Air-gun or Pop-gun: the Fortunes of a Philosophical Instrument," 221–246, in Lissa Roberts, Simon Schaffer, and Peter Dear, eds., *The Mindful Hand: Inquiry and Invention from the Late Renaissance to Early Industrialisation* (Amsterdam: Koninkliijke Nederlandse Akademie van Wetenschappen, 2007), and Lissa Roberts, "A World of Wonders, a World of One", in Pamela H Smith and Paula Findlen, eds. *Merchants and Marvels: Commerce, Science and Art in Early Modern Europe* (New York: Routledge, 2002), demonstrate the power of these spaces for natural philosophical and mathematical conversation and debate well into the eighteenth century.

Chapter 5
Technologies of Pow(d)er: Military Mathematical Practitioners' Strategies and Self-Presentation

Steven A. Walton

Abstract The category of "military mathematical practitioners" consists of those active soldiers and engineers who consciously broadcast their use of mathematical methods to achieve their goals in warfare. These are but a subset of mathematical practitioners more broadly, and they existed on a continuum from the practical to the theoretical, with each demonstrating a mix of the two. In this military case, I investigate the concerns in gunnery and fortification of Thomas Harriot, William Bourne, Thomas Digges, and Edmund Parker—an early-modern scientist, noted author on the mathematical arts, military administrator and author, and a polymath soldier and gunner, respectively—each of whom adopted a certain "mathematical posture" to distinguish themselves in these pursuits. Framed by the work of E.G.R. Taylor, Edgar Zilsel, and Erving Goffman, the examination of how mathematics were actually used by these military mathematical practitioners (which should not be conflated with their actual utility, which is shown here to be often quite lacking) demonstrates the relationship, often a gulf, between theory and practice in one area of the mathematics in later sixteenth-century England. The context, audience, method of development, instruments, and mode of presentation (print vs. manuscript vs. rhetoric) of the mathematical methods applied to warfare also provide evidence of how mathematics was both used and understood as useful in this period to build a self-image of competence and professionalism.

On 28 May 1602 Edmund Parker, a gunner for Queen Elizabeth, died after a brief illness near the Irish Abbey of Bantry, Co. Cork. Sir George Carew specifically mentioned in a letter to Lord Mountjoy that "it hath pleased God to lay his Cross upon us, as I have lost the best cannonier in my opinion that England these many

---

S.A. Walton (✉)
Michigan Technological University, Houghton, MI, USA
e-mail: sawalton@mtu.edu

L.B. Cormack et al. (eds.), *Mathematical Practitioners and the Transformation of Natural Knowledge in Early Modern Europe*, Studies in History and Philosophy of Science 45, DOI 10.1007/978-3-319-49430-2_5

years hath bread."[1] Given that Parker was not noble, wealthy, famous, or otherwise apparently noteworthy, we might pass by this mention as one detail in a field report from a commander to the Lord Lieutenant, but Parker's death figured prominently in the correspondence between Mountjoy and Carew for a number of months. In early May Mountjoy had specifically requested "one parker, an engineer" for his imminent Ulster campaign: He knew him to be "very sufficient and industrious, and fit to do anything wherein we shall have occasion to use him," suggesting that Parker was highly regarded throughout Elizabeth's Irish forces.[2] Parker's death, however, made his transfer to Lord Mountjoy a moot point, and in informing him of that, Carew said that "his loss is no small grief unto me, being an ancient servant of my own, and the want of him will be a great impediment to the service, being (if my judgment fail not) the best cannionier that served her majesty."[3] Mountjoy agreed with the assessment and subsequently requested Carew's other master gunner, Jollye. Carew steadfastly refused to send the other man, "for Canonier or other Atificer (skilfull in the mountures of Ordnance) he had none,"[4] which shows how important it was for a field commander to have an expert gunner on his crew.[5] It appears that when Parker died Carew retrieved his manuscript gunnery notebook, as it still resides in the Carew papers at Lambeth Palace in London.[6] Parker's expertise in gunnery outlived him and helped fuel Carew's personal interest in that art, but the question remains how it is that a low-ranking cannonier could hold the attention of commanders to such an extent.

Episodes such as this demonstrate the importance of mathematically-trained technical personnel to leaders in the early modern world. In the areas of navigation, surveying, gunnery, fortification, and even carpentry, self-conscious mathematical action permeated the thought and action of nobles, tradesmen, and dilettantes. Since the publication of E.G.R. Taylor's monumental *The Mathematical Practitioners of Tudor and Stuart England* in 1954,[7] the study of men who worked with and studied mathematics at the turn of the seventeenth century has been a remarkably stable

[1]Carew to Mountjoy, 28 May 1602 [J.S. Brewer and William Bullen (eds.), *Calendar of the Carew Manuscripts Preserved in the Archiepiscopal Library at Lambeth* [hereafter, *CCML*], (London: Longmans, Green, Reader, & Dyer, 1867–73), vol. 4 (1601–1603), 239 (no. 241)].

[2]H.J. Todd, *A Catalogue of the Archiepiscopal Manuscripts in the Library at Lambeth Palace* (London: Law and Gilbert, 1812), 123, referring to London, Lambeth Palace Library, MS 615, fol. 594; Mountjoy to Carew, 3 May 1602 [*CCML*, vol. 4 (1601–1603), 233–4 (no. 234)]. See also C. Falls, *Elizabeth's Irish Wars* (London: Methuen, 1950), 324–28.

[3]Carew to Mountjoy, 1 June 1602 [*CCML*, vol. 4 (1601–1603), 242 (no. 242)].

[4]Thomas Stafford, *Pacata Hibernia, Ireland Appeased and Reduced* (London: A[ugustine] M[athewes], 1633], 45 quoted in W.A. McComish, "The Survival of the Irish Castle in an Age of Cannon," *The Irish Sword* 9 (1969): 16–21 at 18.

[5]See Mountjoy to Carew, 9 June and 29 July 1602 [*CCML*, vol. 4 (1601–1603), 245 (no. 248) and 285 (no. 274)].

[6]London, Lambeth Palace, MS 280 [hereafter, simply "Parker, *Notebook*"].

[7]E.G.R. Taylor, *The Mathematical Practitioners of Tudor and Stuart England* (Cambridge: Institute of Navigation, 1954).

field. Few new practitioners have been added, and rarely has the category itself been seriously challenged. Most studies of mathematical practitioners have emphasized that this group of men in the later sixteenth century capitalized on the relatively new developments in the apparent mathematization of arts such as navigation, surveying, and cartography in order to build a self-image of competence and—some have even claimed—professionalism. In this literature, the inclusion of military men dealing in troop mustering, gunnery, and fortification has followed.

Taylor's formulation sought to bring to our attention the practitioners who used applied or "mixed" mathematics, as distinct from those scholars and natural philosophers who studied and theorized about "*the* mathematics", and who were both numerous and important for the development of these arts. In this she was aiding the then relatively new "Zilsel Thesis" that saw craftsmen as just, if not more important than natural philosophers for the Scientific Revolution.[8] More recent work has fleshed out how important mathematics began to be in scholarly and even polite discourse in the second half of the sixteenth century. This newly-elevated field of knowledge then helped generate practical and quasi-practical (that is, grounded in crafts but functionally *im*practical) fields upon fields of inquiry.[9]

To be sure, there were grades of practitioners; the complexity of the mathematical skills needed for, say, wine-gauging was far less than that for transoceanic navigation. However, the Taylor formulation that they were all "mathematical practitioners" makes little distinction among these grades to the effect that even relatively humble numerical trades that had existed for centuries such as carpentry could be elevated, at least in the historian's gaze, to a level of importance in some ways possibly undeserved. It is not merely the historian, however, who sought to elevate these men, for the practitioners themselves sought patronage based upon their claim of mathematical competence and many published their arts in the new mathematical framework in order to elevate it. They adopted what I would call a "mathematical posture" to identify themselves. By the late sixteenth century, the self-conceptions (as seen through their activities as well as their public and especially their personal writings) of mathematical practitioners who worked in the military sphere offer a chance to examine what it meant to those actors to be mathematical practitioners. Ultimately they were being consciously "mathematical"

[8]See Edgar Zilsel, *The Social Origins of Modern Science*, ed., Diederick Raven *et al.*, Boston Studies in the Philosophy of Science 200 (Dordrecht: Springer, 2000); A.C. Keller, "Zilsel, the Artisans, and the Idea of Progress in the Renaissance," *Journal of the History of Ideas* 11.2 (1950): 235–240; A. Rupert Hall, "The Scholar and the Craftsman in the Scientific Revolution," in *Critical Problems in the History of Science*, ed. Marshall Clagett (Madison: University of Wisconsin Press, 1959), 3–23; and most recently, Pamela O. Long, *Artisan/Practitioners and the Rise of the New Sciences, 1400–1600* (Corvallis, Ore.: Oregon State University Press, 2011).

[9]See Robin Elaine Rider, "Early Modern Mathematics in Print," in *Non-Verbal Communication in Science Prior to 1900*, ed. Renato G. Mazzolini (Firenze: L.S. Olschki, 1993), 91–113; Katie Taylor, "Reconstructing Vernacular Mathematics: the Case of Thomas Hood's Sector," *Early Science and Medicine* 18.1-2 (2013): 153–179; and Kathryn James, "Reading Numbers in Early Modern England," *BSHM Bulletin* 26.1 (2011): 1–16.

but we do them, and ourselves, a disservice to map *our* modern understanding of being mathematical onto their activities.

Historians have often used a later seventeenth-century or eighteenth-century model of military mathematics and presumed that the sixteenth-century predecessors were engaged in a similar enterprise. While not denying that these earlier military men were both consciously engaging with some mathematics as well as putting some mathematics to use in the field, it is important to understand how sixteenth-century practitioners understood what they were doing. Mathematical knowledge lies on a continuum:

THEORETICAL—PRACTICAL—MATERIAL.

It is worth asking where different types of mathematical practitioners fit on this scale, and further where they fit in different situations. Theoretical knowledge might embody John Napier developing logarithms as he investigated the fundamental behavior of numbers, while material knowledge embodied basic use of numbers to weigh fruit. Practical knowledge—and the core of Taylor's idea of the mathematical practitioner—is when individuals draw from either end of the continuum and when they seek to make application of theory, or more often, to theorize (what I call elsewhere in this chapter, to "mathematize") their material knowledge.

5.1 Military Mathematical Practitioners in Later Sixteenth-Century England

When gunpowder artillery changed the face of warfare in the sixteenth century, first seen in fortification design and then later in naval and eventually field tactics, the practitioners also changed. Historians of warfare uncritically apply modern categories to the sixteenth century as if it were the twentieth: "once Elizabeth was forced to desert diplomacy for warfare, mathematicians and physicists, in cooperation with experienced gunners, developed scientific methods for increasing the effectiveness of artillery, publishing their findings in short, readable, and often well-illustrated textbooks."[10] This improper conflation of gunners with physicists (who we must understand at the time as natural philosophers), as well as attaching the word "scientific" to their methods, imposes far too much mid-twentieth-century baggage to be useful. Period authors certainly did include the military arts as mathematical—John Dee included artillery in his prescriptive taxonomy of mathematics under "menadrie", or, "how Natures Vertue... and force, may be

[10]Henry J. Webb, *Elizabethan Military Science: The Books and the Practice* (Madison: University of Wisconsin Press, 1965), 145. Taylor gets closer, saying more circumspectly that "heights and distances, maps and plans, were decidedly relevant in warfare carried on by artillery" (*Mathematical Practitioners*, 8), but she, too, allows Cold War conceptions of scientific warfare to color her later analysis.

multiplied: and so to ... cast fro, any multiplied ... Vertue, Waight, or Force,"[11]—but those "well-illustrated textbooks" were in fact not relevant for mathematical military *practice.* Let the category of "military mathematical practitioners", then, define those active soldiers and engineers who consciously broadcast their use of mathematical methods to achieve their goals. These men can be found working with artillery and fortifications, which will form the core of this examination, but also in troop ordering and military surveying.[12]

By the eighteenth century, highly trained, technically and scientifically competent corps of artillerymen plied their mathematized art across the battlefields of Europe and across the globe with great skill (in rhetoric if not always battery), furthering our modern characterization of gunnery as a science.[13] They could, it was said in a mechanistic metaphor of the day, reduce a fortress like clockwork.[14] However, before the Thirty Year's War, a close examination of gunnery practice does not bear out this simple story. Later sixteenth-century Englishmen working on both military and mathematical topics did try to use the latter to inform the former in ballistics and in fortification. For example, Thomas Harriot, the "English Galileo" (who, it is worth remembering, also worked on both these topics),[15] shines forth as an eminent mathematician and natural philosopher occasionally interested in military questions. On the other hand, the completely unknown Edmund Parker was the exception that proved the rule in that he was a humble, (probably) non-university educated military engineer who seems to have excelled at mathematics and tried to apply it to gunnery. In between there is William Bourne, self-promoter

[11]John Dee, *The Mathematicall Praeface to Elements of Geometrie of Euclid of Megara* (London: Iohn Daye, 1570), sig. d.i.v and 'Groundplat'. Dee rather typically for the *Praeface* failed to develop exactly *how* guns were mathematical as part of *menandrie* and they and their projectiles were interestingly *not* part of his conception of the study of motion.

[12]In general, see Webb, *Elizabethan Military Science*, and on this last area, see William T. Lynch, "Surveying and the Cromwellian Reconquest of Ireland," in *Instrumental in War: Science, Research, and Instruments Between Knowledge and the World*, ed. Steven A. Walton, History of Warfare 28 (Leiden, 2005), 47–84.

[13]For the later story, see Brett D. Steele, "Muskets and Pendulums: Benjamin Robins, Leonhard Euler, and the Ballistics Revolution," *Technology and Culture* 35.2 (1994): 348–82; Janis Langins, *Conserving the Enlightenment: French Military Engineering from Vauban to the Revolution* (Cambridge, Mass.: MIT Press, 2004); and Ken Alder, *Engineering the Revolution: Arms and Enlightenment in France, 1763–1815* (Princeton: Princeton University Press, 1997).

[14]For the metaphor of the "clockwork siege" ascribed to the later seventeenth-century French siege engineer Vauban, see Jamel Ostwald, "Like Clockwork? Clausewitzian Friction and the Scientific Siege in the Age of Vauban," in *Instrumental in War*, ed. Walton, 85–117.

[15]Matthias Schemmel, "Thomas Harriot as an English Galileo: the Force of Shared Knowledge in Early Modern Mechanics," *Bulletin of the Society for Renaissance Studies* 21.1 (2003): 1–10 and *ibid.*, *The English Galileo: Thomas Harriot's Work on Motion as an Example of Preclassical Mechanics* (Dordrecht: Springer, 2008). See also Matteo Valleriani, *Galileo Engineer* (Dordrecht: Springer, 2010); and Jürgen Renn and Matteo Valleriani, "Galileo and the Challenge of the Arsenal," *Nuncius* 16.2 (2001): 481–503.

(in the sense of the sociological theory)[16] *par excellence*, who published numerous books on the practical mathematical arts, though his proposed mathematical reforms seem to have done little to change the English military system.

Bourne clearly defined a career in the mathematics for himself, publishing almanacs and books on surveying, navigation, and gunnery. He was as far as we know, not university educated and his mathematics appears to have been self-taught or learned at the quayside. As the son of a prominent landowner in Gravesend, Bourne circulated in many spheres throughout his life, acting at times as port-reeve, mayor, jurist, possibly as an innkeeper, and as a volunteer gunner. His attempts to get patronage, however, suggest that he was either within the naval and military sphere for much of his career, or at least wanted to be. The dedications of his works are to Edward Clinton, earl of Lincoln, Lord High Admiral; Sir William Winter, master of the queen's ordnance;[17] and Ambrose Dudley, earl of Warwick, general of the ordnance. Bourne clearly understood the potential for patronage in the military bureaucracy of Elizabethan England, although he never apparently benefited from it more than in his position as gunner in the relatively minor, if strategically important, blockhouse at Gravesend.

Bourne's strictly military publications shed light on the political situation that influenced—but did not necessarily produce—the rise of the military mathematical practitioner. Bourne's *Arte of Shootinge in Great Ordenance* appeared in two editions in the sixteenth century, 1578 and 1587.[18] Bibliographers and historians have repeatedly missed the early edition, leaving the impression that the *Arte* was a response to the Spanish scare of 1587 that resulted in the famed Armada the next year, and hence understood it as a technical response to a political situation. Given the existence of a 1578 print edition, and that fine manuscripts copies were

[16] Erving Goffman, *The Presentation of Self in Everyday Life* (Garden City, NY: Doubleday, 1959), 18: "Social life is described as a multi-staged drama in which people act out different roles in different social arenas depending on the nature of the situation, their particular roles in it, and the makeup of the audience."

[17] See G.L'E. Turner, "Bourne, William (c.1535–1582)," in *Oxford Dictionary of National Biography*, 3rd edition (Oxford: Oxford University Press, 2004) [hereafter '*Oxford DNB*'] and the older but still useful, E.G.R. Taylor, "William Bourne: A Chapter in Tudor Geography," *The Geographical Journal* 72.4 (1928): 329–339. Turner claims without clear evidence that Bourne learned gunnery from Winter.

[18] The canonical Maurice J.D. Cockle, *A Bibliography of Military Books up to 1642*, 2nd edition (London: Holland Press, 1957), no. 35 and Webb, *Elizabethan Military Science*, both missed the first edition. The unique 1578 copy, held at the Royal Artillery Institution (STC 3419.7), is set with different type than the 1587 copies (italic vs. black letter) and has hand-drawn or pasted-in illustrations, although the layout and catch-words agree in both editions. The Stationer's Company transcript records the *Arte* as licensed to Henry Bynneman on 22 July 1578 and Borne noted in his *An Almanacke and Prognostication for x. yeeres* (1581), that his "booke called the Art of Shooting in great Ordenaunce" was already in print [E.G.R. Taylor (ed.), *A Regiment for the Sea and Other Writings on Navigation by William Bourne*, Hakluyt Society 2nd ser. 121 (London: Cambridge University Press, 1963), 328]. Further, John Dee had a copy of the 1578 edition in his library, possibly a gift from Bourne himself [J. Roberts and A.G. Watson (eds.), *John Dee's Library Catalogue* (London: Bibliographical Society, 1990), 37].

in circulation from at least 1572, it is evident that the *Arte of Shootinge* was created for a different audience. Rather than being a practical manual (although it is that), a 1572 presentation copy of the *Arte* seems more likely to have been a ploy on Bourne's part to curry favor with Lord Burghley or the Ordnance Office after an intensely local, if prosperous, career in Gravesend.[19] The lavish copy of the *Arte of Shootinge* was in fact made *before* any of Bourne's successful navigation books appeared, suggesting that for a volunteer gunner who had no contacts at Court, gunnery presented itself as the topic to garner courtly attention (That is was with navigation that Bourne seems to have actually succeeded should also give us pause about claiming too much for the publishing success of the military arts in Tudor England).

The illustrations in the different editions of Bourne's *Arte of Shooting* suggest the changing audience expected for the manuscript versus print editions. Since all versions have the same content but presented in slightly different orders and manners, the presentation of the material suggests differing strategies. In the manuscript, for example, a diagram of cannonball trajectories includes the label on the vertical portion, "the perpendicular line of fall of the shot" while the woodcut in the print edition omits the label.[20] The information is instead relegated to the text. Although this may be partly a function of the medium (pen vs. woodcut), it is also indicative of a more visual, demonstrative approach to the manuscript medium, or the vocabulary projected at the recipient ('perpendicular' is a technical term in geometry and local motion comes from natural philosophy, whereas 'straight down' might be more vernacular). This approach can be seen in the more elaborate depiction of quadrants for inclined and declined shooting in note 16 of the manuscript. There, "the quadrant . for down the hill" and "the quadrant . for up the hill" are more carefully separated and delineated, placing the emphasis on the instrument, not the action.[21] This strategy of an "instrumental" approach is quite characteristic of the self-presentation of the gunners in defining their role to the State, for they worked diligently to advertise the potential uses of their art, as embodied by their instruments, rather than trying to demonstrate their individual skill in actually performing tasks.[22]

By comparison, one other important contemporaneous military mathematical practitioner worth noting, a man surprisingly understudied in this mathematical con-

[19]London, British Library, MS Sloane 3651. See also Taylor, *Regiment for the Sea*, 441–2. The 1587 edition is virtually identical to the 1578 edition, and they are expanded from the manuscript versions, which also provided some of the substance for Bourne's *Inventions and Devices* and his *Treasure for Travellers* (both also 1578).

[20]Bourne, *Arte of Shooting in Great Ordnaunce* (1578; London: [Thomas Dawson] for Thomas Woodcocke, 1587), 40 and MS Sloane 3651, fol. 25r.

[21]Bourne, *Arte of Shooting*, 86 and MS Sloane 3651, fol. 36v.

[22]The distinction is much like that between Leonardo, whose letter to the Sforzas enumerated what practical works he could personally make *for* them (canals, engines, painting), versus Galileo's *Le operazioni del compasso geometrico et militare* (Padua, 1606), which advertises to purchasers what could be done *with* his military compass.

text although he actually had military experience, is Thomas Digges. Educated in mathematics by John Dee and published with his father in solid geometry and astronomy by the 1570s,[23] Digges' personal calling was civil service and he acted as MP (Wallingford) before issuing a call for mathematical reform in both navigation and military matters. Serving then as muster-master and fortifications inspector in the Netherlands from 1585 to 1588, Digges soon thereafter reissued enlarged versions of his 1570s publications, incorporating at least some of his experiences from the Dutch Wars, and leading Robert Norton to later refer to him as "that rare Souldier and Mathematician."[24]

However, when one considers the content of Digges' mathematical military books, it becomes evident that the years spent in actually performing military duties did not increase his ability to apply mathematics to war. The use of mathematics in troop ordering, calculation of garrison strength, and other logistical matters remains rudimentary from the 1579 to 1590 editions of *Stratioticos*, although he does increase the number of examples of how to calculate troop sizes, and his section on gunnery begins to address questions that later ballisticians (as well as natural philosophers) will find of interest. In the first edition, Digges asks "Certeine Questions in the Arte of Artillerie, by Mathematical Science joyned with Experience, to be debated and discussed" (pp. 181–91; sig. Z.iij–[&.iv]). These questions included simple proportional calculation questions ("The *Proportion* found by *experience* in one *Peece* of the different *Randges* of Iron and Lead Bullets make, whether the same proportion hold in any other *Peece* longer or shorter, shooting the same *Bullets*, whatsoeuer hir length be"), more complex algebraic questions ("If two *Peeces* of the same *Length* & *Bullet* be charged with one kinde of *Pouder*, but seueral *Waights*, I demande whether the *Randges* shalbe proportional to the said weights, or to the [square], [cube], or [fourth] *Rootes* of the said waights"), and complex, if confusing, trigonometric inquiries ("If the quantitie of the *Cone* of euery *Peece* proportionallie charged, be by experience found, I demande whether

[23]Thomas Digges, *Alaæ seu scalæ mathematicæ* (London: Thomas Marsh, 1573), on astronomy; Leonard and Thomas Digges, *A Geometrical Practise, named Pantometria* (London: Henrie Bynneman, 1571), on plane and solid geometry and reissued in 1591 with a new section of artillery definitions; and Leonard and Thomas Digges, *An Arithmeticall Militare Treatise, named Stratioticos* (London: Henrie Bynneman, 1579), expanded and reissued in 1590.

[24]Robert Norton, *Of the Art of Great Artillery* (London: Edw. Allde for Iohn Tap, 1624), sig. A2r-v. See in general, Stephen Johnston, "Digges, Thomas (c.1546–1595)," in *Oxford DNB*. Of particular relevance here is A.R. Hall, *Ballistics in the Seventeenth Century: a Study in the Relations of Science and War with Reference Principally to England* (Cambridge: Cambridge University Press, 1952), 43–49, which generally overestimates Digges' importance to the field; Stephen Johnston, "Making Mathematical Practice: Gentlemen, Practitioners and Artisans in Elizabethan England," Ph.D. dissertation, Cambridge University, 1994; Johnston, "Like Father, like Son? John Dee, Thomas Digges and the Identity of the Mathematician," in *John Dee: Interdisciplinary Studies in English Renaissance Thought*, ed. Stephen Clucas, International Archives of the History of Ideas 193 (Dordrecht: Springer, 2006), 65–84. Eric H. Ash, "A Perfect and an Absolute Work: Expertise, Authority, and the Rebuilding of Dover Harbor, 1579–1583," *Technology and Culture* 41.2 (2000): 239–268, notes Digges' extensive service in this particular aspect of his life, but understandably does not examine the military connections or his foreign service.

then this *Elipsis* shal not make an *Angle* with the *Parabola Section* equal to the distance betweene the grade of Randon proportioned, and the grade of the vttermost *Randon*"). No attempt, however, is made to answer any of these fifty-three questions, although a concluding section does refer disparagingly to earlier mathematical and geometrical attempts at ballistics by Daniel Santbech, Girolamo Ruscelli, and Niccolo Tartaglia.[25]

By the 1590 edition of *Stratioticos*, Digges could only manage to reprint the same questions and commentary verbatim, "to give parcticioñers some Encouragement to try Conclusions" and add perfunctory marginal notes to the questions that simply said "yes", "no", or "not always"—and even this he could manage for only 32 of the 53 the questions posed eleven years before.[26] (Fig. 5.1) Although he claimed that this was a prolegomenon to a larger work on "Martiall Pyrotechnie and great Artillerie, hereafter to be published", he admitted that "there are yet many *Mysteries* that by farther pro[o]fes, and trials *Experimental*, I must resolve, before I can reduce that *Art* to suche perfection as can content me.... [M]y first endeavours shal be entierly to finishe the *Treatise* of that newe *Science* of manedging this newe furious Engine & rare *Invention* of great *Artillerie.*" Nothing of this longer promised work is known to survive, if indeed it was ever composed.[27] Digges, then, clearly conceived of gunnery as a mathematical discipline, and did work to frame the analysis of the practice in the most complex mathematics of the day. The results of this work, however, are less than impressive from a practical point of view and point to the effective failure of mathematics in the sixteenth century to accurately describe ballistic trajectories or the relationship between all the variables in a cannon shot.[28] It is not surprising, then, that contemporaneous and later military

[25]Daniel Santbech, "De Artificio Eiaculandi Sphaeras Tormentarias" in his *Problematum astronomicorum et geometricorum sectiones septem* (Basel: Henrichum Petri et Petrum Pernam, 1561); Girolamo Ruscelli, *Precetti della militia moderna... tutta l'arte del Bombardiero* (Venice: Marchiò Sessa, 1568); Nicolò Tartaglia, *Nova Scientia* (Venice: Stephano da Sabio, 1537) and *Quesiti et Invenzioni Diverse* (Venice: Venturino Ruffinelli, 1546 and 1554) which were later epitomized by Cyprian Lucar as *Three Bookes of Colloquies Concerning the Arte of Shooting in Great and Small Peeces of Artillerie* (London: Thomas Dawson for Iohn Harrison, 1588) and *A Treatise Named Lucarsolace* (London: Richard Field for Iohn Harrison, 1590). See also Matteo Valleriani, *Metallurgy, Ballistics and Epistemic Instruments the* Nova scientia *of Nicolò Tartaglia* (Berlin: Edition Open Access, 2013) and Raffaele Pisano and Danilo Capecchi, *Tartaglia's Science of Weights and Mechanics in the Sixteenth Century: Selections from* Quesiti et inventioni diverse: *Books VII–VIII*, History of Mechanism and Machine Science 28 (Dordrecht: Springer, 2016), 39–86 on ballistics and fortificaiton.

[26]*Stratioticos* (1590), 349–60; quote at 361.

[27]*Pantometria* (1591), title page and *Stratioticos* (1579), sig. [&.iv], respectively, italics in the original. Digges died in 1595, but Thomas Smith returned to Digges' questions in his *The Arte of Gunnerie... by Arithmeticke Skill to be Accomplished* (London: Richard Field for William Ponsonby, 1600); he had little success in furthering answers, focusing instead on less complex things that could, as his subtitle announced, "by arithmeticke skill ... be accomplished."

[28]Even E.G.R. Taylor *begins* her book noting that "at the opening of the eighteenth century the many technical difficulties inherent in making and using accurate and reliable instruments and apparatus, not to speak of finding correct theoretical formulae, were unsolved and insoluble until

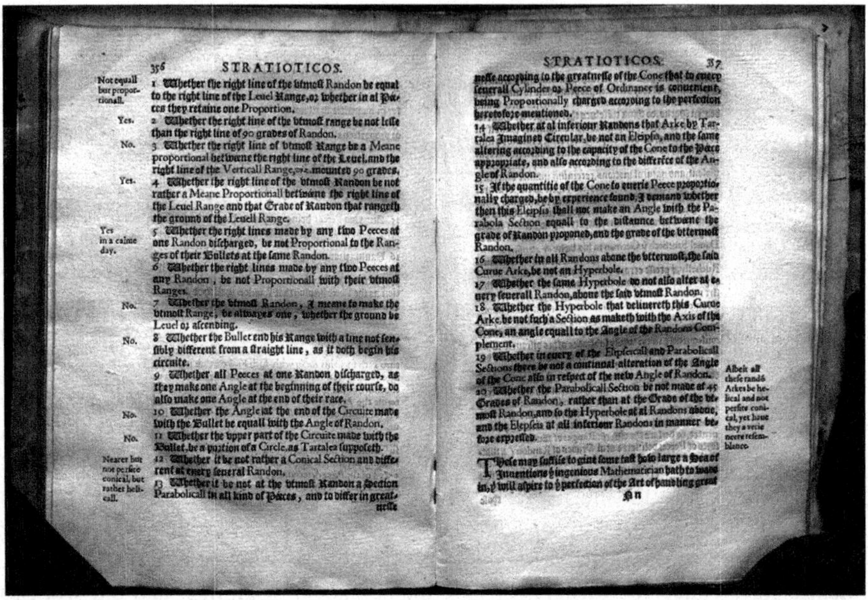

Fig. 5.1 Leonard and Thomas Digges, *An Arithmeticall Warlike Treatise Named Stratioticos* (London: Imprinted by Richard Field, 1590), 356–357. (By courtesy of the Department of Special Collections, Memorial Library, University of Wisconsin-Madison)

mathematical practitioners in gunnery tended to drop Digges' line of reasoning in favor of more tractable approaches.

Digges himself serves as a sort of "boundary object"[29] between the educated world of natural philosophy and the practicing world of the armed services. He attended University College, Oxford, although he did not graduate, and then he found a "gentlemanly career of service" in governmental service. His attempts to fuse mathematical theory and artillery practice bore little fruit, although in terms of what he presented artillery to be—a very mathematical art, even if his work on

some further advance... had been made" (*Mathematical Practitioners* [note 7, above], 3). The full description of a cannonball's flight is not a closed analytic function, but must be determined experimentally. The earliest effective mathematical ballistic gunnery handbooks date to the late nineteenth century: James M. Ingalls, *Exterior Ballistics* (Fort Monroe, Va.: U.S. Artillery School, 1885) and *Exterior Ballistics in Plane of Fire* (New York: D. Van Nostrand, 1886); Lawrence L. Bruff, *Exterior Ballistics, Gun Construction, and U.S. Seacoast Guns* (West Point, NY: United States Military Academy Press, 1892).

[29]On boundary objects, see Susan Leigh Star, "The Structure of Ill-Structured Solutions: Boundary Objects and Heterogeneous Distributed Problem Solving," in *Distributed Artificial Intelligence*, ed. Les Gasser and N. Huhns (London: Pitman, 1989), II: 37–54 and Star and James Griesemer, "Institutional Ecology, 'Translations' and Boundary Objects: Amateurs and Professionals in Berkeley's Museum of Vertebrate Zoology, 1907–39," *Social Studies of Science* 19.3 (1989): 387–420 at 393.

it remained rather qualitative—his work demands further investigation. But even more enlightening in the matter of "technologies of pow(d)er" are two other military mathematical practitioners: one university educated, one not; one with direct noble support, the other without. In the end, these two men offer a glimpse of where working military mathematical practitioners situated themselves in late Elizabethan society and what they did when they found themselves there.

5.2 Military Mathematical Practice

One of the principle difficulties of earlier studies of the "military mathematical practitioners" (and indeed all mathematical practitioners) has been a reliance on sources that do not necessarily reflect what they were doing and thinking, but rather that reflect what others thought of them or how they presented their field to others. E.G.R. Taylor and Henry Webb, for example, relied heavily on the printed record, which gave a broad, yet bounded, view of the mathematical practitioners' world. Taylor, for example, missed numerous sixteenth-century fortification engineers working in England, precisely because works on fortification where not published until well into the seventeenth century.[30] Surveys like A.R. Hall's *Ballistics in the Seventeenth Century* ignored writings and notes of the military mathematical practitioners themselves. Rarely have scholars tried to understand what the mathematical practitioners themselves studied and worked on, as distinct from how they sought to present themselves to the outside world, which may not be entirely the same. Personal notebooks, journals, or field notes of the military mathematical practitioners encode content of the profession as they thought it to be, and demonstrate the activities these men learned (although not necessarily ever actually performed). More importantly, they demonstrate how military mathematical practitioners as a group argued for their own importance.[31]

5.2.1 *Fortifying*

Fortification is one side of military mathematics, and the one most apparently tied to formal mathematics at this time. In the late fifteenth century, Italian engineers began experimenting with various improvements to castle design that better responded,

[30]Talor, *Mathematical Practitioners*; Horst de la Croix, "The Literature on Fortification in Renaissance Italy," *Technology and Culture* 4.1 (1963): 30–50; and Barbara Donagan, "Halcyon Days and the Literature of War: England's Military Education before 1642," *Past & Present* 147 (1995): 65–100.

[31]For a parallel sort of unconscious self-presentation by personal writing, see Nicholas Popper, "The English *Polydaedali*: How Gabriel Harvey Read Late Tudor London," *Journal of the History of Ideas* 66.3 (2005): 351–381.

they believed, to the new, relatively powerful if inaccurate cannon fire beginning to pervade warfare. Their solution was two-fold: first, bring the high, medieval walls down to low, earth-filled ramparts that were harder to hit and could reflect or absorb incoming fire; second, redesign the layout of the castle walls along polygonal lines that made it far harder for an attacking force to get close to the walls, at least not without coming under withering crossfire. Although the medieval castle entailed little mathematical theory or understanding, the Renaissance fortress evolved into an idealized, symmetrical, and above all geometrical construction.[32] By the middle of the sixteenth century, mathematicians, too, began conceiving of fortification *as* a geometric problem. Giacomo Lanteri, for example, was the first author to treat fortification design as a purely abstract, geometrical problem in his *Due dialoghi... del modo di disegnare le piante delle fortezze* (1557), which he specifically advertised as being "according to Euclid."[33] (Fig. 5.2) An ideal late-sixteenth century plan of a fortification, then, began with a regular polygon (if the landscape would allow it, irregular otherwise), and then designed flanks, bastions, gorges and all the other elements of the new system from geometrical principles of lines of fire and attack. Pioneered in Italy, this new *trace italienne* fortification arrived in England in the 1550s and 1560s through the arrival of Italian fortification engineers in the employ of Henry VIII and Elizabeth, and the written work of Robert Corneweyle, Peter Whitehorne, and Jacopo Aconcio directed at state leaders like William Cecil later Lord Burghley, and various nobles such as Bedford and Northumberland.[34] Although only one major English fortress was built in this style in this period (Berwick-upon-Tweed), many smaller ones superficially adopted the

[32]J.R. Hale, *Renaissance Fortification: Art or Engineering?* (London: Thames and Hudson, 1977) provides the best brief introduction. For the early history, see Gianni Perbellini, *The Fortress of Nicosia, Prototype of European Renaissance Military Architecture* (Nicosia: Anastasios G. Leventis Foundation, 1994) and Pietro C Marani, *Disegni di fortificazioni da Leonardo a Michelangelo* (Firenze: Cantini edizioni d'arte, 1984).

[33]Giacomo Lanteri, *Due dialoghi di M. Iacomo de' Lanteri....: ne i quali s'introduce Messer Girolamo Cantanio... & Messer Francesco Treuisi... 'a ragionare Del modo di disegnare le piante delle fortezze secondo Euclide; et Del modo di comporre i modelli & torre in disegno le piante delle citt'a* (Venetia: Costantini, 1557). Curiously, Lanteri's teacher of mathematics was Girolamo Cataneo, who broke from strictly symmetric geometrical constructions and allowed for irregularities due to terrain and local conditions in his own *Opera nuovo di fortificare* (Bresica: Gio. Battista Bozola, 1564); see Horst de la Croix, "Literature on Fortification," 40–41.

[34]See Robert Corneweyle, *The Maner of Fortification of Cities, Townes, Castelles and Other Places, 1559* (Richmond, Surrey: Gregg, 1972); Lynn White Jr., "Jacopo Acontio as an Engineer," *American Historical Review* 72.2 (1967): 425–444; and my "State Building through Building for the State: Domestic and Foreign Expertise in Tudor Fortifications," in *Expertise and the Early Modern State*, ed. Eric Ash, Osiris 25 (2010): 66–84. Aconcio's "lost" book on fortification, originally written in Italian and possibly translated into Latin, survives in an English translation by Thomas Blundeville (another mathematical practitioner) that was independently rediscovered by Stephen Johnston (Oxford University), by me, and by Paola Giacomoni (Università di Trento); see Paola Giacomoni (ed.), *Jacopo Aconcio: Trattato sulle fortificazioni*, Istituto Nazionale di Studi sul Rinascimento, Studi e Testi 48 (Firenze: Leo S. Olschki, 2011). My modernized transcription will appear in the journal *Fort* from the Fortress Study Group in 2017.

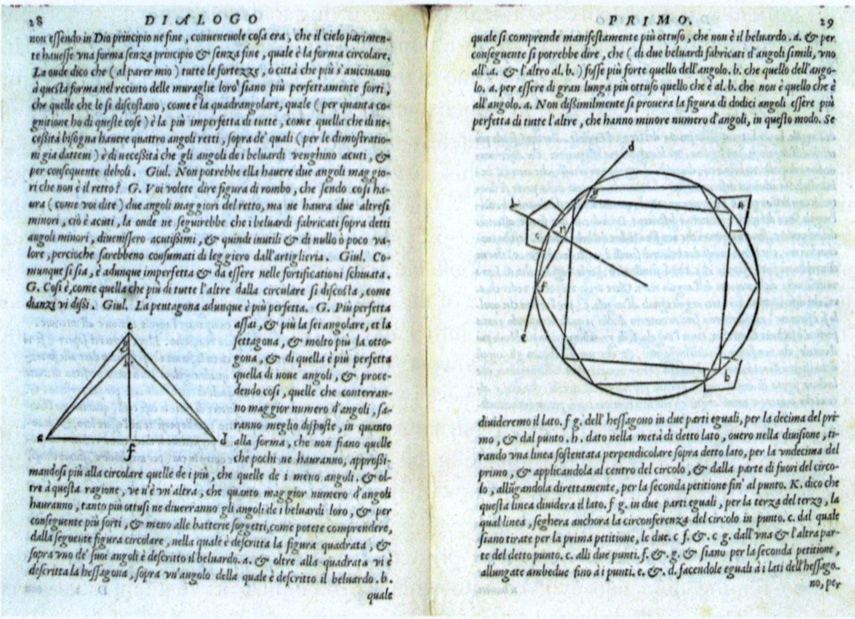

28 DIALOGO

non essendo in Dio principio ne fine, conuenevole cosa era, che il cielo parimente hauesse vna forma senza principio & senza fine, quale è la forma circolare. La onde dico che (al parer mio) tutte le fortezze, ò città che più s'auicinano à questa forma nel recinto delle muraglie loro siano più perfettamente forti, che quelle che le si discostano, come è la quadrangolare, quale (per quanta cognitione ho di queste cose) è la più imperfetta di tutte, come quella che di necessità bisogna hauere quattro angoli retti, sopra de' quali (per le dimostrationi gia dattemi) è di necessità che gli angoli de i beluardi venghino acuti, & per consequente deboli. Giul. Non potrebbe ella hauere due angoli maggiori che non è il retto? G. Voi volete dire figura di rombo, che sendo cosi haura (come voi dite) due angoli maggiori del retto, ma ne haura due altresi minori, ciò è acuti, la onde ne seguirebbe che i beluardi fabricati sopra detti angoli minori, diuenissero acutissimi, & quindi inutili & di nullo ò poco valore, percioche sarebbeno consumati di leggiero dall'artiglieria. Giul. Comunque si sia, è adunque imperfetta & da essere nelle fortificationi schiuata. G. Cosi è, come quella che più di tutte l'altre dalla circulare si discosta, come dianzi vi dissi. Giul. La pentagona adunque è più perfetta. G. Più perfetta assai, & più la sei angolare, et la settagona, & molto più la ottogona, & di quella è più perfetta quella di noue angoli, & procedendo cosi, quelle che conterranno maggior numero d'angoli, saranno meglio disposte, in quanto alla forma, che non siano quelle che pochi ne hauranno, approssimandosi più alla circolare quelle de i più, che quelle de i meno angoli. & oltre à questa ragione, ve n'è vn'altra, che quanto maggior numero d'angoli hauranno, tanto più ottusi ne diuerranno gli angoli de i beluardi loro, & per conseguente più forti, & meno alle batterie soggetti, come potete comprendere, dalla seguente figura circolare, nella quale è descritta la figura quadrata, & sopra vno de' suoi angoli è descritto il beluardo. a. & oltre alla quadrata vi è descritta la hessagona, sopra vn'angolo della quale è descritto il beluardo. b. quale

PRIMO. 29

quale si comprende manifestamente più ottuso, che non è il beluardo. a. & per conseguente si potrebbe dire, che (di due beluardi fabricati d'angoli simili, vno all'. a. & l'altro al. b.) fosse più forte quello dell'angolo. b. che quello dell'angolo. a. per essere di gran lunga più ottuso quello che è al. b. che non è quello che è all'angolo. a. Non dissimilmente si prouera la figura di dodici angoli essere più perfetta di tutte l'altre, che hanno minore numero d'angoli, in questo modo. Se diuideremo il lato. f g. dell' hessagono in due parti eguali, per la decima del primo, & dal punto. h. dato nella metà di detto lato, ouero nella diuisione, tirando vna linea sostentata perpendicolare sopra detto lato, per la vndecima del primo, & applicandola al centro del circolo, & dalla parte di fuori del circolo, allũgandola direttamente, per la seconda petitione fin' al punto. K. dico che questa linea diuidera il lato. f g. in due parti eguali, per la terza del terzo, la qual linea, segherà anchora la circonferenza del circolo in punto. c. dal quale siano tirate per la prima petitione, le due. c f. & . c g. dall'vna & l'altra parte del detto punto. c. alli due punti. f. & . g. & siano per la seconda petitione, allungate ambedue fino à i punti. e. & . d. facendole eguali à i lati dell'hessagono, per

Fig. 5.2 Giacomo Lanteri, *Due dialoghi... 'a ragionare Del modo di disegnare le piante delle fortezze secondo Euclide* (Venetia: Appresso Vincenzo Valgrisi & Baldessar Contantini, 1557), 28–29. (Used with permission from Eberly Family Special Collections Library, Penn State University Libraries)

style and by the 1570s it was *de rigeur* to understand this new geometric method of fortification design.

In this matter, three contemporary Englishmen represent relatively distinct types of military practitioners who all nonetheless speak to the growing sense of mathematics as being crucial to fortifying. In the work of Thomas Harriot, Edmund Parker, and Richard Norwood we see over the space of a half-century a succession of a man theorizing about the components of the system, a practical autodidact judging the relevance of the new system, and a pedagogical geometer proselytizing to the masses, respectively. These roles are, of course, neither mutually exclusive nor a necessary historical progression, although the development does underlie the diffusion of an awareness of mathematical practice in fortification.

Thomas Harriot approached fortification in his capacity as mathematics tutor to Henry Percy, the Ninth Earl of Northumberland and more specifically, possibly as tutor to Prince Henry Stuart.[35] As the most 'academic' of the military mathematical practitioners in this study, it is not surprising that Harriot came at the study of the

35Others have argued that Harriot did his military work (at least the gunnery) for Walter Raleigh and hence for practical reasons; I argue otherwise in *Thomas Harriot's Ballistics and English Renaissance Warfare*, Occasional Paper no. 30 (Durham: Durham Thomas Harriot Seminar, 1999).

trace italienne in a formally Euclidean manner, trying to derive typical proportions for the elements of the bastion in an attempted proposition–theorem format. His work is atypical for the field at this time in that it is in Latin and because it is trying to derive a theory of fortification rather than its practical (numeric) application of it, and his work seems uncharacteristically derivative. Perhaps it is not surprising, then, that as far as we can tell from his surviving manuscripts, Harriot abandoned his attempt after only a few folios.[36] On the other hand, Edmund Parker—the gunner mentioned in the introduction—provides a practical construction method for a fortification in his manuscript notebook. He notes that for "the curtain a .100. yards devided into .8. parts, . . . the lengthe of the bulwark [is] 2 of these parts," and that "the breadth [of the bulwark] divided into 3 and 1/3 [of these is] for the breadth and length of the casemate."[37] In Parker's work, there is no theoretical framework, just simple proportional rules of thumb. These two approaches between the Earl's or the prince's study and the Irish trenches strike us as reasonable and appropriate for the two venues, but it was also reasonable in the other direction: these two practitioners operated in these modes because of how they wished or needed to be seen *in* those venues.

By the time fortification entered print in England, the standard presentation method for the field had evolved into what we recognize as mathematical story problems. They were designed to teach relationships and mathematical operations more than they necessarily were to teach the subject matter at hand.[38] In his 1639 treatise on military architecture, for example, Richard Norwood posed the reader the following problem: "There is a heptangular Fort, whose Gorge-line is 14. rods, the flanke 12. rods, and the curtain 38. rods: I demand the quantity of the other parts of such a septangular fort, the flanked angle of its bulworke being 79 2/7 degrees?"[39] Any pretense to a real-world application that takes into account terrain or cost is absent. To calculate the proportions of the new seven-sided fortification, a reader would have to understand the relationship between the gorge, flank, and the curtain (the width of the neck of a bastion, its face, and the wall between the bastions, respectively) and then apply rules of proportion (the ubiquitous 'Rule of Three' that stated for a simple proportion given three knowns, calculate the unknown, as in $a : b :: c : d$) learned in previous chapters or from earlier mathematical study. By the early seventeenth century, the *trace italienne* fortification theory had become one standard method of instruction in applied geometry, or the "mixt mathematicals." Here we see the difference between a "framing" approach based upon theory by Harriot, a

[36]His fortification pages are London, British Library, MS Add. 6788, fol. 55–65. They bear a similarity to Samuel Marolois', *Opera mathematica, ou Oeuvres mathématiques traictans de géométrie, perspective, architecture et fortification* (Hagæ-Comitis: Henrici Hondii, 1613–14), so is either a late foray by Harriot (†1621), or his work prefigures Marlois.

[37]"Rules touching Great Ordnance" in Parker, *Notebook* (note 6, above), esp. fol. 32v–36.

[38]Wolff-Michael Roth, "Where is the Context in Contextual Word Problems?" *Cognition and Instruction* 14.4 (1996) 487–527.

[39]Richard Norwood, *Fortification or Architecture Military* (London: Tho. Cotes for Andrew Crooke, 1639), p. 59.

"constructing" approach based upon simple field-ready methods by Parker, and a "calculating" approach based upon the ability to do simple arithmetic and the Rule of Three in Norwood.

Fortification engineers would have had little traction in a society that was not directly interested in the objects and knowledge they were peddling, whether built or unbuilt. For both the fortifiers and the gunners that we will take up shortly, the material culture of forts and diagrams, rulers and dividers, and cannon and instruments all hold meaning that both legitimated and reinforced the status of the practitioners. Both directly and indirectly, these elements are of considerable importance in the general study of mathematical practitioners, and especially for late Tudor and early Stuart military mathematical practitioners, who were producing new knowledge in the marriage of mathematics, instruments, and military hardware. They then tried to deploy this new knowledge (or the semblance of it) to gain status in the Elizabethan and Jacobean state, and in the end it came to define them as persons in a nascent professionalism.

Erving Goffman argued that "the meaning of an object (or act) is a product of social definition and that this definition emerges from the object's role in society at large." While for "smaller circles" the role of those objects/acts might be a given (that is, among a local and specialized group of practitioners, tools of the trade are taken for granted), for a wider audience, their definition "can be modified, but not totally re-created."[40] These objects can fluctuate between crucial functional tools and stage props (think, for example, how the military mace became the royal scepter) and take on an important symbolic position in the negotiation of roles.

5.2.2 Gunning

Gunnery forms another part of the category of military mathematical practitioners. Modern historiography has fused gunnery onto the other mathematical arts such as astronomy, navigation, and surveying. Ballistics, of course, defines the revolution in mechanics epitomized by Galileo and modern artillery ranging certainly does rely on precision measurement and calculation. Renaissance gunners *were* taught mathematics, to be sure, but the question has rarely been asked what mathematics the gunner was taught, or more importantly actually needed or could use to perform contemporary practice of operating heavy artillery. Period gunnery textbooks were at pains to emphasize that the gunner needed to know his arithmetic, and the principle mathematics taught was the Rule of Three and the rules of right triangles

---

[40]Erving Goffman, *Frame Analysis: An Essay on the Organization of Experience* (New York: Harper and Row, 1974), 39.

(though not the Pythagorean Theorem).[41] More advanced instruction might include very limited trigonometry, but this would usually end at similar triangles.

Many treatises are also designed around story problems, much as Norwood's fortification example above. In Thomas Smith's *The Art of Gunnery* (1600), problems are primarily framed in this manner: "How by Arithmeticke skill you may know how with one and the self like charge in powder and bullet, how much farre or shorte any peece of Ordinance will shoot, in mounting or dismounting her any degree ... at any degree of the randon."[42] Smith provides a mathematically-derived (though faulty) solution. Many other problems are of the form, "Question: If a Saker at point blanke convey her bullet 200 paces, and at the best of the randon shoot 900 paces, what will that Cannon do which at point blancke shoots 360 paces?"[43] Smith here relies on the Rule of Three to derive his solution that the ultimate range of a Cannon is 1630 paces, *i.e.*, using the simple proportion:

$$\frac{200}{900} : \frac{360}{1630}$$

Ballistic trajectories and ranges do not follow simple proportional scaling relationships and since Smith makes no distinction for the different sizes or materials (and hence density) of the cannonballs or any external factors such as wind, his answer bears little relation to what gunners actually encounter. For Smith and other authors, simple mathematical rules were the answer, and in making it so, they presented artillery *as* mathematical.

The manuscript record for gunnery practice and instruction, however, reveals a different picture. Mathematics in these sources are extremely limited beyond enumeration, with no Euclidean postulates or proto-algebraic formulations. Gunnery instruction math was very simple, cookbook-style mathematics.[44] For example, when an anonymous writer penned a "secrets of gunmen" manuscript in the mid-sixteenth century, it was clear that higher mathematics was not on his mind. He

[41] See for example, Digges and Digges, *Stratioticos* (note 23, above), ch. 1–9, pp. 1–52; Smith, *The Arte of Gunnery* (note 27, above), 1–8; and Robert Norton, *The Gunner Shewing the Whole Practise of Artillery* (London: A.M. for Humphrey Robinson, 1628), 1–30, which sets out theorems for artillery practice.

[42] Smith, *Arte of Gunnery*, 46. "Randon" is the farthest range of a shot, and describes great speed, force, or violence [*OED*, s.v. "Random 1a"] and hence distance. The modern meaning of *random* as haphazard or without exactness [*OED*, s.v. "Random 3"] seems to derive from the tendency of things that rush headlong with great speed and violence to loose accuracy.

[43] Smith, *Arte of Gunnery*, p. 35. "Saker" and "cannon" are proper names for different sizes of ordnance, of nominally 3½ and 7 or 8 inches bore diameter, respectively.

[44] The didactic style of the manuscript suggest use in the academies; see Steven A. Walton, "The Bishopsgate Artillery Garden and the First English Ordnance School," *Journal of the Ordnance Society* 15 (2003): 41–51, and "Proto-Scientific Revolution or Cookbook Science? Early Gunnery Manuals in the Craft Treatise Tradition," Ricardo Cordoba (ed.), *Craft Treatises and Handbooks: The Dissemination of Technical Knowledge in the Middle Ages*, De Diversis Artibus 91 (Turnhout: Brepols, 2013), 221–236.

opened the treatise by noting that "First you must know good salt-peter from bad", noting various blunders a gunner might make and then how "to make good powder thereof for all manner of good shot." Only third does a very basic mathematical understanding appear: "you must know all peeces *measurably* to charge or lade them, & to parte [the]m over field, land, or roades, to shoote as hereafter declareth."[45] In a chapter devoted to "Certayne Questions Arithmeticall" for the master of the ordnance, Thomas Digges asks just three questions about cannon—on weight, on charge, and on range, all of which he solves by the Rule of Three—and then adds the topics he would have covered, had space allowed:

> The weight, quantitie, and number, of Powder, Shotte, and sundrie sortes of Ordinaunce to bee used at a *Batterie*: howe to mounte all sortes of Peeces, to strike anye marke at *Randon*: the number of Carriadges, of Ladles, Rammers, Scouters, Waddes, Tampions, Cartages, Matches, Barrels, or Lastes of Poweder &c. Also, the number of Gunners, Assistantes, Pioners, Smythes, Carpenters, and others *Artificers*, to attende on the *Altillerie*, what number of Horses and Oxen to drawe them, the wayght of all sortes of Peeces, the charges of them, theyr Wheeles and Carriages.[46]

Even if a tiny nod to ballistic gunnery does appear ("howe to strike anye marke at Randon"), it is immediately overwhelmed by the myriad numeration tasks that count as "Arithmeticall" to Digges—and recall that these are the very matters a muster-master like Digges would need to care about. Even if print authors sometimes develop elaborate mathematics lessons for their readers, those lessons are entirely absent in the manuscript notebook tradition. If we take the latter as a proxy at least *closer* to the practice of the gunners themselves (an axiom of this study), then gunners themselves begin to look less mathematical than modern historiography implies, though they certainly remain numerical. Ultimately, this is in distinction to other mathematical practitioners, who increasingly deployed more and more complex mathematics (both observations and calculations) to develop their art.[47]

The key component of a gunners' education lay quite explicitly in a cookbook understanding of materials: "you must know how to make 3 or 4 sortes of fireworks at least, whether it be by water or land, *if you will get lords wages*."[48] Patrons might possibly be interested in mathematical gunnery, but fireworks, it would seem, were the real selling point for gunners. Gunners were tasked to provide them for festivals and royal activities, and it is fireworks and gunpowder recipes that fill page after page of the manuscript notebooks. This observation can also be made for the printed works relating to gunnery. The very first such English book, Peter Whitehorne's *Certain Waies for Ordering Soldiours in Battleray* (1560), translates Machiavelli's *Art of War* and then adds a new, extended appendix on gunpowder, fireworks, and

---

[45]Oxford, Bodleian Library, MS Ashmole 343 [hereafter, "Secrets of Gunmen"], fol. 128r, emphasis added. Although the manuscript is an early seventeenth-century copy (suggestive of durability of interest in such topics), the material is clearly from the mid- to late sixteenth century.

[46]Digges, *Stratioticos*, 66.

[47]*E.g.*, Susan Rose, "Mathematics and the Art of Navigation: the Advance of Scientific Seamanship in Elizabethan England," *Transactions of the Royal Historical Society* 14 (2004): 175–184.

[48]"Secrets of Gunmen," fol. 128r, emphasis added.

fortification. Cyprian Lucar's influential *Three Bookes of Colloquies concerning the Arte of Shooting* (1588)—an adaptation of books I-III of Niccolò Tartaglia's *Quesiti et Inventioni Diverse* (1546)—included a compilation of military matters from "divers good authors in diverse languages" considerably longer than the translation itself, with a large section on fireworks. Thus, when gunners approached gunning in this period, they appear to have done so mathematically—but only in the same way a cook could be considered to approach a recipe "mathematically". These men were learning mathematics, and using it to some degree, but the contemporary evidence suggests that their definition of being "mathematical" was not as "scientific" as ours, and further that their deployment of mathematics was not as extensive or as complex as we tend to assume.

Edmund Parker, whose death was recounted at the opening of this essay, offers a glimpse of a working gunner who was literate enough to leave us a record of his work, and a record that appears to have for his own use, rather than any composition intended for other eyes. That Sir George Carew specifically retrieved Parker's notebook shows its importance, but it has no pretense to being a developed textbook or presentations copy, like Bourne's manuscripts. Parker was not a gentleman nor even a Captain as far as we can tell, and yet both Carew and Lord Mountjoy considered him "the best cannionier that served her majesty."[49] Parker is perhaps uncommon as a gunner in that he had a wide knowledge of mathematical authors, but more importantly, his impressive notebook reveals what applying "the mathematics" to gunnery looked like: *quantifying* gunnery, yes; *theorizing* about it, no.

The notebook gives us a wonderful glimpse into the working life of a military mathematical practitioner. Opening with a decidedly non-military recipe for fishing—using bread, cheese, *aqavite*, and honey thrown into the water in "small pieces"—and also used from both ends with page after unrelated page of jottings, notes, and calculations, the notebook is roughly one-third gunnery, one-third navigation, and one-third general rules of proportion and mathematics. Some of the pages are clearly textbook-like examples we might still see today, as for example for calculating the time it would take for a cistern to drain, given a certain sized stopcock. Most of the pages, however, do not deal directly with the mathematics we would expect from a cannonier, leaving the impressions that Parker considered himself mathematical, but gunnery itself was another matter.

Parker refers to works by other mathematical practitioners such as William Bourne, Robert Recorde, and John Blagrave, so he is clearly reading the mathematical literature. Further, he is quite advanced in his mathematics, dealing with 'surds' (irrationals) and he seems adept in his proto-algebraic manipulations. Throughout the manuscript, Parker is always performing calculations. But these are virtually always the simplest calculations—such as the extraction of roots or the common Rule of Three—to which virtually all of the mathematics in military use reduce.

[49]Carew to Mountjoy, 1 June 1602 [*CCML* (note 1, above), vol. 4 (1601–1603), 242 (no. 242)].

Parker *never* uses his mathematics for the practice of gunnery as it is usually understood or assumed by modern historians, that is, for ranges and trajectories. The one place he does use it is in the calculation of "experimental gunnery" (his words) where he figures out how many arrows of a certain diameter could be shot from a given cannon. His Rule of Three calculations suggest that a falconet two inches in diameter would take 5 sheaf arrows, and a demi-culverin of four inches would take one dozen one-inch diameter arrows, while an eight-and-a-half-inch diameter mortar holds 200 small arrows, 80 musket arrows, or 50 one-inch arrows, and so on.[50] This is in effect simply calculating how many small circles fit into a bigger circle—which is mathematically interesting—but hardly militarily useful. Ultimately he dubs this the "new art of archery", and although he claims to have shown its efficacy in Ireland, the idea seems to have died with him.[51] In another place Parker carefully marks out a twelve-inch circle, calculates the area, and then finds sub-circles that were one-half, one-third, and one-sixth the area of the circle, the sum of their areas being that of the large circle itself. Again, this is wonderfully mathematical, technically deft, but not at all useful militarily. It has no relevant application in gunning, or even in navigation or surveying, the other mathematical arts Parker explores in his notebook.

The areas of experience that Parker seems interested in, at least as indicated by what he chose to record, and their contribution to his persona as a military mathematical practitioner—both elements of self-definition, I would argue—are in the realm of instruments. He and other military mathematical practitioners clearly allied themselves to mathematical instruments as a certain badge of honor. One of the key issues in gunning was not how to place the shot on target, for that was often done at point blank range and rarely needed much aiming, much less instrumental precision, but rather on figuring out the imperfections in specific artillery pieces so that they could be aimed accurately. Thus, up to the eighteenth century when manufacturing allowed much better quality control over casting and finishing, gunners had to determine whether or not the bore of the barrel was centered in the cannon and whether the axes of the bore and the exterior were collinear. When they were not, gunners had to compensate in sighting their shots. This also determined how they "disported" the piece, that is, mounted temporary, supplemental sights onto the barrel as aiming aids. The most common way to determine this was to measure the wall thickness of the cannon all the way around and see if one side was thinner or thicker than another. This, however, is not a perfectly simple operation.

Published sources offered their own instruments to make dispart measurements. In 1545, Niccólo Tartaglia presented a dialogue between himself and a gunfounder who wished to know how to tell if the bore be in the "midst of the Mettall."[52] He

[50] Parker, *Notebook* (note 6, above), fol. 19 and 25v.

[51] Parker, *Notebook*, fol. 83.

[52] Lucar, *Three Bookes of Colloquies* (note 25, above), 43–5 (bk. I, colloquy 23), which is apparently from Tartaglia's *Quesiti*, as it is not in the *Nuova Scienza* (see Valleriani, note 25, above).

self-importantly describes an instrument of his invention to accomplish the task, while the gunfounder asks him to "Thinke a while of it, for I have asked this doubt of many ... Engenars, and have not found any of them able to resolve mee therein." Tartaglia proposed an assembly of two long parallel pieces of wood a "brace"[*i.e.*, *braccia*] longer than the length of the bore, connected by two shorter crosspieces, "somewhat more than half the thicknes of the Peece at the tail." His illustrations indicate that the shorter pieces fasten the longer ones parallel to one another after the fashion of a tic-tac-toe grid at one end. Then, by inserting one of the longer pieces into the bore and pressing it tight to one side or the other and measuring the gap between the other piece of wood and the outside of the gun at various places around the circumference, the gunner may be able to tell if the bore is indeed coaxial with the outside of the piece. In 1578, William Bourne described Tartaglia's grid instrument and also offered another one: an oversized, double-ended caliper.[53] Two pieces of wood, "double the length of the hollow or concavity of the piece," are fixed on a hinge in their center allow the gunner to insert one leg into the bore and close that end on the cannon wall, the thickness of which will register as the gap at the other end of the instrument. Interestingly, Borne does not actively critique Tartaglia as much as simply add another alternative.

Edmund Parker, however, actively engaged in the debate.[54] He offered another instrument: take, he says, a 6–7 foot ladle staff, "verye drye and as straight as maye be" and put two rammer heads for the particular cannon you wish to test on it, one at the end, and one three or four feet up the shaft. The rammer heads were to be made "a littell to lowe and sumthinge tapringe towardes the for[e] end." Wool or cloth strips were then nailed or glued to the heads to allow them to snugly fit the bore. Two or three feet outside the mouth, the gunner placed a three-inch thick block of wood, turned "verye true and rounde" and "all wayes ... verye ner as brodde as the mettell at the mouth." Then, with a string attached to a nail driven into it, the block is moved in or out on the rammer until the string just grazes the cannon at the tail and the mouth. Then, if the string also grazes fore and aft all the way around the piece, it is truly bored; if not, then "marke the differenc ther of whiche showethe how muche it is." Parker clearly invests a great deal of effort here and sets himself apart with his active engagement with the printed sources. In concluding these instructions, Parker writes,

> This I hould far better than tartaglies or bornes device for that ther instrument beinge so longe and so weacke will swaige to and frowe wher by no gret sertentye may be geven to it ther resen is good if the instrument wer as good but this other waye is withe out all dout if the guner worke Cuningelye or eles blame the workeman and not the device .1596.

Without a doubt, then, Parker knew Bourne's works and both knew of Tartaglia's (Parker presumably read Bourne's books, but whether each of the Enligshmen read Tartaglia in the Italian is uncelar; transmission through intermediaries is also possible). Parker rightly notes that these instruments would be prone to "swaige" or

[53]Bourne, *Arte of Shooting*, 8–9.

[54]The following is from Parker, *Notebook*, fol. 11.

racking since the shortest ones to test falcons would need to be at least six or seven feet long, and one of Bourne's second type to measure a cannon would be at least 18 ft long.

Two years later, Parker also invented a range-finding instrument. He writes in his notebook, "suppose the thing you desire to know the distance to—a castle or a tower or anything else—but it must be either higher or lower than the ground you shall stand on."[55] His rangefinderhad sights and a silk plumb line to level it, and used a scale graduated from zero to 75 degrees by fives. To use it, the gunner took the angle of elevation or depression at one point, then walked a certain distance either towards or away from the target and took another elevation/depression. He does not note explicitly in the notebook how one then takes these two angles and a distance and converts to the overall range (quite easy with trigonometry, but that was not likely available to Parker or his contemporaries, so it must have relied on ratios of chords), yet he feels the need to remind the reader (and perhaps himself) that noting the number where the plumb line lies means reading it off the scale. This invention was clearly important to him, as it is dated 1598 and is the only signed page in the entire manuscript.[56] That he takes possession of the instrument is shown in his explicit term "*my* instrument", and indeed on another dated page, he speaks of a variation he had previously invented: "*my* quadrant .1596."[57]

It is with instruments, ultimately, that gunners and other military mathematical practitioners linked their work to mathematics.[58] Edmund Parker invented instruments. Cyprian Lucar, when expanding Tartaglia (who only mentions the *squadra*), describes numerous instruments and paid to have the costly woodcuts made to illustrate them (or at least John Harrison, his publisher, recognized that illustrations would sell the book).[59] Gunnery textbooks describe them in detail—Thomas Smith even included an oversized foldout plate to illustrate a near life-sized quadrant.[60] Instruments of brass and wood have been found in shipwrecks like the *Mary Rose* and in other military contexts showing that both master gunners and common gunners frequently used them in practice.[61] Elaborate gilt instruments were

---

[55] Parker, *Notebook*, fol. 29.

[56] It may not be a signature as much as an ascription of invention as it is in a clear italic hand while the majority of the MS is a rather scrawling secretary hand.

[57] Parker, *Notebook*, fol. 10.

[58] See Steven A. Walton, "Mathematical Instruments and the Creation of the Scientific Military Gentleman," in *Instrumental in War*, ed. Walton (note 12, above) 17–46.

[59] Harrison also published other mathematical works, such as Robert Recorde, *The Pathewaie to Knowledge Containyng the First Principles of Geometrie* (London: J. Kingston for Ihon Harrison, 1574), which relied heavily on illustrations.

[60] Smith, *The Arte of Gunnery* (note 27, above), facing 58.

[61] Kurt Petersen, *Det Militære Målesystem: Kaliberstokken Og Dens Udvikling Fra 1540 Til 1850* (Lyngby, Denmark: Polyteknisk Forlag, 2005); A. Konstam, "A Gunner's Rule from the 'Bronze Bell' Wreck, Tal-y-Bont," *Journal of the Ordnance Society* 1 (1989): 23–26; Ruth R. Brown, "Comment on The Tal-y-Bont Gunner's Rule," *Journal of the Ordnance Society* 2 (1990): 71–2; Jeremy N. Green, "Further Information on Gunner's Rules or Tally Sticks," *Journal of the*

made for the princely audience, which suggests both a market for such expensive instruments and that the more humble practitioners did manage to convince the elites that instruments were symbolic for gunners and defining of mathematical gunnery itself, even if the gunners gained little status bump from doing so. Bourne's *Arte of Shooting*, rather than explain how one uses arithmetic or geometry to calculate ranges, provided a table to convert the inch readings from a practitioner's rule to desired ranges, suggesting that it was expected that the practitioners would buy or make their own rules.[62] And if we need a mental image of a practitioner making the explicit statement, "I am a gunner and I am mathematical", we need look no further than Richard Wright, who in the early 1560s, drew a self-portrait in his gunner's manual where he held aloft not a sword or dagger or lintstock, but a gunners' rule as his baton of office.[63] (Fig. 5.3)

Thus, in promoting themselves based upon instruments, and with the backing of the instrument-making community, the military mathematical practitioners made *words* out of *things*. That is, they used objects to signify the knowledge necessary (or implied) to control those gunpowder technologies and turned them into technologies of pow(d)er. They also argued for the power inherent in the instruments themselves and they turned the physical objects into social objects which emerge as "signifying symbols... that can be used for meaningful communication"—a conversation we can still hear today.[64] But even more so they made their gunners' rules, levels, quadrants, and other instruments not into status objects, which convey prestige (typically economic) based upon their quality, but rather as "esteem

Ordnance Society 2 (1990): 25–32; Winifred Glover, "The Spanish Armada Wrecks of Ireland," in *Excavating Ships of War*, ed. Mensun Bound (Oswestry, Shrops.: Nelson, 1998), 51–63; David S. Weaver, "The English Gunner's Caliper," *Arms Collecting* 33 (1995): 111–25; Colin Martin, "De-Particularizing the Particular: Approaches to the Investigation of Well-Documented Post-Medieval Shipwrecks," *World Archaeology* 32 (2001): 383–99; Alex Hildred, "The Material Culture of the *Mary Rose* (1545) as Fighting Vessel: The Uses of Wood," in *Artefacts from Wrecks: Dated Assemblages from the Late Middle Ages to the Industrial Revolution*, ed. Mark Redknap (Oxford: Oxbow Books, 1997), 51–72; and Alexzandra Hildred, *Weapons of Warre: The Armaments of the Mary Rose* (Portsmouth: Mary Rose Trust, 2011), 392–407.

[62]Incidentally, it also suggests that gunners need not have been able to do the calculations themselves. Michael Korey, *The Geometry of Power: Mathematical Instruments and Princely Mechanics around 1600* (Munich: Deutscher Kunstverlag, 2007), 19, puts it nicely by noting that a table or instrument is "less than a calculator than... an instrument for *avoiding* calculation," emphasis original.

[63]London, Society of Antiquaries, MS 94 [hereafter, "Wright, *Notes*"]. As early as the late sixteenth century, gunners took the invention of instruments as a both signs of accomplishment as well as keys to employment and advance: consider the *Radio Latino* invented by Latino Orsini that could be used for gunnery, fortification, and surveying; the quadrant of Johann Carl, *Zeugmaster* and engineer of Nurenberg; or the instrument described by Thomas Bedwell in his *Aurea Regula Coss, Nova Geometrica* ("The Golden Algebra, a New Geometry"). For all of these, see Walton, "Mathematical Instruments and the Creation of the Scientific Military Gentleman".

[64]R.S. Perinbanayagam, "How to do Self with Things," in *Beyond Goffman: Studies on Communication, Institution, and Social Interaction,* ed. S.H. Riggins (Berlin: Mouton de Gruyter, 1990), 315–340.

Fig. 5.3 Richard Wright Self-Portrait, from his *Notes on Gunnery*, Society of Antiquaries, London, MS 94, fol. 2 (© The Society of Antiquaries of London)

objects," which "show how well a person fulfills general duties irrespective of rank."[65] Since English military mathematical practitioners never saw great gains in status due to their skills, it is a testament to their strategies in the late sixteenth and early seventeenth centuries that we conceive of gunnery and fortification as more scientific than, say, wine-gauging or carpentry, even though both of those arts also deployed instruments, calculation, and recipes (formulæ) as well.

5.3 Conclusion: The Rise of the Military Mathematical Practitioner

It seems clear from surviving documents such as gunners' notebooks—manuscripts not apparently designed as formal treatises, publications, or presentation pieces—that geometry, instruments, measuring, and recipes (especially fireworks) were seen as keys to self-development and definition. When the technical ideas upon which the new military technologies of gunnery and fortification began to appear in didactic

[65]S.H. Riggins, "The Power of Things: the Role of Domestic Objects in the Presentation of Self," in Riggins (ed.), *Beyond Goffman*, 347–49. They also then became *occupational objects*, which include both objects that credential for the common gunners (using wooden gunners' rules and gauges), as well denote privilege and prestige for the princes (with their complex, compound gilt instruments in their *Kunstkammern*).

print treatises, they were received as branches of the mathematics. That these treatises may not have been supremely useful for the practical undertaking of any of these arts is relatively unimportant; the textbook-like nature of the print treatises reinforces the idea that general mathematical rules could be devised and deployed for military action, and by extension, for society more widely conceived. Military commanders increasingly saw military activities of gunnery, fortification, troop mustering, and provisioning as mathematical and the new polygonal fortification style obviated any real questioning of this assumption. Elites also invested, literally, in buying fine mathematical instruments for military tasks. Whether they used them at all is another question entirely.[66]

All that remains here, then, is to consider how—and how well—the military mathematical practitioners promoted their skills as mathematicians in order to secure patronage, obtain jobs, define the field, and advance the art itself. Their overall success in all these strategies varies greatly. By the end of the seventeenth century there was very little change in England for gunners: they had become regular employees of the Ordnance Office or of town corporation and their position in the hierarchy leveled off or even declined slightly by the time of the Civil War.[67] The position of fortification engineer had been rare and relatively prestigious in the sixteenth century and one which England mostly asked Continental "experts" to fill, but with the need for a great number of field fortifications during the Civil War, design and construction along geometrical lines became a matter of general instruction for trained bands in many cities.[68] Both mathematical technologies became to some extent gentlemanly pursuits—gentlemen often owned books on both gunnery and fortification, prints of fortification design, and fine instruments for both—but only the fortification engineers became marginally elite, the study being taught in academies and by royal tutors as the proper role of the aristocracy.

The same cannot quite be said about the practice of gunnery. For gunners, objects became a defining element of self-presentation, even if others were not fully ready to accept the new self-definition of the military mathematical practitioner. This accounts for a surprising *lack* of attention paid to these actors by their audience (the State) and explains the failure of the military mathematical practitioners to gain any

[66]A. J. Turner, "Mathematical Instruments and the Education of Gentlemen," *Annals of Science* 30 (1973): 51–88; Gerard L'E. Turner, *Elizabethan Instrument Makers: The Origins of the London Trade in Precision Instrument Makers* (New York: Oxford University Press, 2001); and especially Jim Bennett, "Early Modern Mathematical Instruments" *Isis* 102.4 (2011): 697–705. Instruments would later become foundational for the mathematics itself: see for example, Hester Higton, "Does Using an Instrument Make You Mathematical? Mathematical Practitioners of the 17th Century," *Endeavour* 25.1 (2001): 18–22.

[67]Richard W. Stewart, *The English Ordnance Office, 1585–1625. A Case Study in Bureaucracy* (Woodbridge: Boydell, 1996).

[68]Practice geometrical entrenchments, for example, seem to have been dug in the Artillery Garden outside Bishopsgate in London during the Civil War; see Steven A. Walton, "The Tower Gunners and the Artillery Company in the Artillery Garden before 1630," *Journal of the Ordnance Society* 18 (2006): 53–66 at 58.

real status in Renaissance England (Mario Biagoli has shown that the Italian case is, as always, rather different)[69] until well into the later seventeenth and eighteenth centuries, by which time they are absorbed into a larger military bureaucratic network of planners and general staff. Socially, gunners became soldiers while fortifiers became architects.[70] This distinction may have been a result of the failure of mathematical ballistic theory to accurately predict artillery fire until at least the late the eighteenth (or, one might argue, the twentieth) century.[71] Nonetheless, in retrospect we can still divine some clear relationships between various theoretical frameworks and these actors.

Early modern technical practitioners employed mathematics to bolster their crafts and to claim social position based upon their skill. The technologies related to military gunpowder weaponry, whether offensive or defensive, appealed to mathematics to do the same. Curiously, the consequence of this strategy was to increasingly remove the creative individual military practitioner from the equation, because mathematical technologies came to be seen as encoding objective facts, and empower the commander who could order his forces like clockwork.[72] Because instruments are supposed to simply record truth (when you read a thermometer, it matters not who or where you are; if it reads 32°, then that is what it is), individual users are constrained by what has been called "artifact physics" (what the objects simply "do").[73] Users are then taken to play passive roles as transmitters of the instruments' knowledge and truth. Although scholars will now agree that scientific knowledge is socially constructed, objects themselves (and especially scientific instruments) remain the core of positivistic behavior, at least for their users. Further, scientific instruments and especially quantitative scientific instruments are taken as

[69] Mario Biagioli, "The Social Status of Italian Mathematicians, 1450–1600," *History of Science* 27 (1989): 41–95 and more recently, Cesare S Maffioli, "A Fruitful Exchange/Conflict: Engineers and Mathematicians in Early Modern Italy," *Annals of Science* 70.2 (2013): 197–228.

[70] For a related study of the rise of architecture, see Anthony Gerbino and Stephen Johnston, *Compass and Rule: Architecture as Mathematical Practice in England 1500–1750* (New Haven, Conn.: Yale University Press 2009).

[71] Fluid mechanics of projectile flight would continue to dog natural philosophers until well past the mid-eighteenth century; see Hall, *Ballistics in the Seventeenth Century* (note 24, above); John F. Guilmartin, Jr., "Ballistics in the Black Powder Era," in *British Naval Armaments*, ed. Robert D. Smith (London: Trustees of the Royal Armouries, 1989), 73–98; and Steele, "Muskets and Pendulums" (note 13, above), and see note 28, above.

[72] This has been contested in modern scholarship, but it is just such scholarship that showed how forcefully seventeenth-century natural philosophers argued for the objectivity of their "facts" at the time. See Mary Poovey, *A History of the Modern Fact: Problems of Knowledge in the Sciences of Wealth and Society* (Chicago: University of Chicago Press, 1998) and Steven Shapin and Simon Schaffer, *Leviathan and the Air-Pump: Hobbes, Boyle, and the Experimental Life* (Princeton: Princeton University Press, 1985) and the literature that has flowed from their work.

[73] The useful term "artifact physics" is from Christopher R. Hoffman and Marcia-Anne Dobres, "Conclusion: Making Material Culture, Making Culture Material," in *The Social Dynamics of Technology: Practice, Politics, and World Views*, ed. Dobres and Hoffman (Washington, DC: Smithsonian Institution Press, 1999), 209–22 at 216. See also Davis Baird, *Thing Knowledge: a Philosophy of Scientific Instruments* (Berkeley: University of California Press, 2004).

uncontestable by many outside the science studies field. Renaissance gunners and fortifiers for all intents and purposes also believed this and acted accordingly.

But of course instrument users make choices all along. Despite some actions being seen as wrong (*e.g.*, using a carpenter's level to order troops) or incompatible (*e.g.*, a microscope cannot view the stars), there is a great deal of latitude to decide what to use for a given purpose, how to use it, and where and when to employ (or deploy) it. Objects in general and instruments in particular are human-made and do not have their own intrinsic natures; we can construct their nature as a negotiation between physical behavior and social agreement. This is *not* an argument for deception on the part of the military mathematical practitioners, even if their claims do often appeal to potential more than their actions demonstrate utility. Through gunners (claiming to be) divining meaning from the numbers they read off wooden and brass instruments, instruments themselves became part of the presentation act, much like the judge's gavel, the warrior's sword, or the magician's wand (the reader may decide which analogy is most apt).

Technology plays a dynamic, performative role in making or breaking relationships between ideas and social groups, and the embeddedness of technologies and techniques (Marcel Mauss' idea of *technique* as a "total social fact") stands against the fact that when we entering new technical relationships throughout our lives, we have the option to reaffirm old identities or outright adopt or manufacture new ones to accommodate the situation.[74] Erving Goffman clearly enunciated the key to the way we may view military mathematical practitioners in this period:

> Society is organized on the principle that any individual who possesses certain social characteristics has a moral right to expect that others will value and treat him in an appropriate way .... [Consequently], an individual who implicitly or explicitly signifies that he has certain social characteristics ought in fact to be what he claims to be. ... [O]thers find, then, that the individual has informed them as to what is and to what they *ought* to see as the 'is'.[75]

The social characteristics in this case are precision, mathematization, and accuracy. And although Goffman would certainly allow a great deal of charlatanism to occur under the guise of the individual "signifying that he has certain social circumstances," no overt chicanery is implied here for the military mathematical practitioners.

Scholars today are not at all surprised by the claim that that "agency... is inscribed onto the material world of resources and power," which then gives "certain individuals control of the objects produced, control of the technologies and technicians involved, control over the value systems that regulate the status

[74] See Bryan Pfaffenerger, "Worlds in the Making: Technological Activities and the Construction of Intersubjective Meanings," in *The Social Dynamics of Technology*, 147–164, as well as the conclusion to that volume, esp. 213–215ff.

[75] Goffman, *Presentation of Self* (note 16, above), 13, emphasis original.

of... technicians, and control of both esoteric and practical knowledge."[76] The gunners and fortifiers presented the objects of gunnery as *matters of concern* to their patrons and inscribed agency into those mathematical objects. They argued through action and demonstration that their patrons should care about mathematics and the instruments in which mathematics was embedded. In the process, these military practitioners turned their own behavior and claims into *matters of fact* (that is, uncontested and reified stable entities free of emotion or social concerns) and turned themselves into what I have called military mathematical practitioners.[77]

In the end, even though these matters of fact could not be sufficiently established—that is, the predictive power of gunner's rules and quadrants could not make artillery fire perfectly accurate until centuries later—and even though the military mathematical practitioners in England could not gather together a sufficient quorum to attain the status to which they aspired, they did not fail in their trying. They bound up their material and practical knowledge, fused it to a small degree with the theory as it existed at the time, and presented themselves as masters of their new field. In that, they produced not just social objects, but complete identities that could be understood by the militarily literate in the early modern period.

[76]Marcia-Anne Dobres, referring to the work of Judith McGraw and Ruth Schwartz-Cowan, among others, in her, "Technology's Links and *Chaînes*: the Processural Unfolding of Technique and Technician," in *The Social Dynamics of Technology*, 129. This claim was made for differences of gender, but the same argument works in reference to difference in ability based upon access to and fluency with mathematics. While Dobres is interested in the *chaîne opératoire* methodology of studying the transformation of raw materials into products and the meanings engendered along the way, her more general point is exactly what I am arguing here for the military mathematical practitioners: "*while* undertaking productive activities, individuals create and localize personal and group identities, making statements about themselves that are 'read' by others with whom they are interacting. Technical acts can thus be treated as a medium for defining, negotiating, and expressing personhood" (129, emphasis in original).

[77]The terms "matters of concern" and "matters of fact" are from Bruno Latour, "Why has Critique Run out of Steam? From Matters of Fact to Matters of Concern," *Critical Inquiry* 30 (2004): 225–48, esp. 246 (and reiterated in his "From Realpolitik to Ding Politik, or How to Make Things Public," in *Making Things Public: Atmospheres of Democracy*, ed. Bruno Latour and Peter Weidbel ([Cambridge, Mass,: MIT Press, 2015], 14–41), although I have inverted his assessment of current critique.

Chapter 6
Machines as Mathematical Instruments

Alex G. Keller

Abstract When Leonardo da Vinci claimed that mechanics is the paradise of the mathematical sciences, he was launching an approach to the design and improvement of machinery. His notebooks witness how far he himself had proceeded in the application of mathematical techniques to such tasks.

Only in the middle years of the sixteenth century, as artillery came to dominate the field of warfare did such ideas begin to take hold. Ancient works on mechanics were studied enthusiastically, like the *Mechanical Problems*, and Archimedes, who became a culture hero. Meanwhile Tartaglia, particularly in his *Nova Scienza*, suggested that similar methods could be used to describe bodies in motion. The development of artillery imposed a new style of fortification, which required a new military engineer, one whose skills were closer to those of the surveyor. Those who wished to hold command in war might need some help with problems of this kind, encouraging mathematicians to devise novel instruments, that would assist in carrying out observation and calculation.

The publications of Agricola and Biringuccio on mining and metallurgy also demonstrated how widely machinery was employed in these lucrative sectors of the economy. Bringing all this together, the first printed books of mechanical invention often illustrated mathematical instruments as well as their machines, while they insisted on the way their machines embodied simple mathematical concepts.

When Leonardo da Vinci (1452–1519) called mechanics "the paradise of mathematics," as it bore the fruit of both those sciences, he was making a connection that sounds as dramatic as it was original.[1] The mental link 'paradise—fruit' recalls to mind an image of the Garden of Eden and its Tree of Knowledge, a striking claim

[1]"la meccanica e il paradiso delle scientie matematiche perche con quella si viene al frutto matematico," from MS E8b in J. P. Richter (ed.), *The Literary Works of Leonardo da Vinci* (London: Phaidon, 1970), II: 241, 1155.

A.G. Keller (✉)
University of Leicester, Leicester, UK
e-mail: agk@leicester.ac.uk

L.B. Cormack et al. (eds.), *Mathematical Practitioners and the Transformation of Natural Knowledge in Early Modern Europe*, Studies in History and Philosophy of Science 45, DOI 10.1007/978-3-319-49430-2_6

that makes this one of the best known and most often cited gnomic remarks that Leonardo jotted down in his notebooks. For most of his contemporaries, the term 'mechanics' would probably suggest the *Mechanical Problems*, a book traditionally then attributed to Aristotle although probably composed somewhat later, perhaps in the third century BCE by a member of the Aristotelian school.[2] In that book, the author tries to explain the working of a number of simple devices by a law of the lever, crudely expressed, though there is no mention of more complex machines, such as mills.

It is true that not much later Archimedes did show the way to apply geometrical methods to physical problems in a manner much more logical and sophisticated than the older work. In short treatises—hardly more than essays—he explained why levers raise heavy objects, why solid bodies placed in water float or sink, and how much water they displace. However, these treatises deal only with statics and were not easily applicable to bodies in motion. For Plutarch, his biographer who lived long after Archimedes, the great mathematician could not really have found intellectual satisfaction in his machines; he could only have meant them to serve to impress the vulgar—either a not-quite-legitimate monarch or the *hoi polloi*—who could not appreciate more abstract ideas. Supposedly for Archimedes and Plato (Plutarch would have assumed), mathematical theorems and proofs dealt with ideal situations and one should not think of them as applicable to real life, which is by necessity so untidy. In Plutarch's eyes only the desperate need of Archimedes' native city of Syracuse could have obliged him to devote his great intellect to devising weapons which could hold at bay the besieging army of Rome.[3] All the same, Plutarch does give an account of the strange engines of war Archimedes designed to defend the city, and so whether or not he intended to, he gave life to the image of Archimedes as the "great inventor".

Consequently, Archimedes became a plausible role model for anyone who believed that they could employ their talents to provide better machines for war or for peace. Later inventors might, for example, try to work out how Archimedes had hauled a ship on to dry land. As screw-pumps were commonly attributed to Archimedes, even sometimes bearing his name, further development with that machine would have added to inventors' prestige in the eyes of the Renaissance. Likewise, because screw-gears (AKA worm-gears), in which a screw engages with a cog whose teeth are cut to fit its threads, also involve the motion of a helix, they seem to have acquired a certain cachet as being more ingeniously 'mathematical' than other gears. The name they are sometimes given, 'endless screw', and the related 'Archimedean screw' doubtless helped as well.

[2]Aristotle, *Minor Works*, ed. and trans. Walter Stanley Hett (Cambridge Mass.: Harvard University Press, 1936); Paul L. Rose and Stillman Drake, "The PseudoAristotelian Questions of Mechanics in Renaissance Culture," *Studies in the Renaissance* 18 (1971): 65–104; Matteo Valleriani, "The Transformation of Aristotle's Mechanical Questions: A Bridge Between the Italian Renaissance Architects and Galileo's First New Science," *Annals of Science* 66.2 (2009): 183–208.

[3]Plutarch, "Life of Marcellus," xiv in Plutarch, *Plutarch Lives: Agesilaus and Pompey, Pelopidas and Marcellus*, trans. Bernadotte Perrin (Cambridge: Harvard University Press, 1917), 5: 470–77.

Even before Leonardo, some scholarly intellectuals had already interested themselves in applied mathematics, which for them were as the "fruit of the mathematical sciences." The applications, however, were methods for solving problems of measurement, surveying land, estimating heights of buildings or differences of level, or in cartography. Angle measurement was obviously an important element in these techniques and the instruments used by astronomers and astrologers to measure celestial angles did not differ greatly from those used for work on the earth's surface. (Indeed, the terms 'mathematician' and 'astrologer' were in common parlance often almost interchangeable).[4] Since navigation required observation of the sky it is hardly surprising that navigational instruments of the sixteenth century often turn out to be modifications of those long-used by astronomers for their own purposes. However, although the architect Brunelleschi is supposed to have delighted in studying and devising the then still relatively novel clocks, and certainly designed new machines for use in the construction of the great dome of Florence Cathedral, neither he nor his immediate successors made explicit the connection that Leonardo was to make.[5]

After all, the art of constructing machines had by then been practiced for many centuries. We could even go back in time and in concept to the chain of pots which probably originated in the Hellenistic Levant. Their function was to transfer the circular motion of a beast of burden into the vertical haul of the endless chain in order to raise water.[6] Not long thereafter, watermills made their appearance by the first century BCE at the latest, and these were the first manufacturing devices to use an inorganic power source and to clearly add the additional complexity of a gearing system, which converts a 'tool' into a machine.[7] In medieval Christendom and Islam, and far beyond, such machines became ubiquitous and were employed in a wide range of industrial uses, particularly in Western Europe. Such mills were constructed according to what became tried-and-true techniques, by craftsmen who for the most part had little interest in changing their methods of work or structures they built; once established, procedures became traditional, and few, if any of the builders seem to have asked how they could be made more efficient, for they accomplished what was needed from them.

Within the rather slim theoretical inheritance from Antiquity that was somewhat amplified by the time of the *quattrocento*, it is possible to trace a concept of five, or in some versions six, simple machines that could be interpreted as reducible

---

[4]See Katherine Hill, "'Juglers or Schollers?': Negotiating the Role of a Mathematical Practitioner," *British Journal for the History of Science* 31 (1998): 253–74.

[5]For Brunelleschi see Frank D. Prager and Gustina Scaglia, *Brunelleschi: Studies of his Technology and Inventions* (Cambridge, Mass.: MIT Press, 1970).

[6]T. Schiøler, *Roman and Islamic Waterwheels* (Odense: Odense University Press, 1973).

[7]An overshot waterwheel for a mill is also described by Vitruvius, *De Architectura* X.5 (hereafter cited as 'Vitruvius'). Terry S. Reynolds, *Stronger than a Hundred Men, a History of the Vertical Waterwheel* (Baltimore: Johns Hopkins University Press, 1983). Also Steven A. Walton, "reCOGnition: Medieval Gearing from Vitruvius to Print," *AVISTA Forum Journal* 19.1/2 (2009): 28–41.

to the fundamental principle of the lever. The leading architectural writer of the *quattrocento*, Leon Battista Alberti, does mention these basic tools in his *De Re Aedificatoria* (1443–52). A copy in the Fitzwilliam Museum in Cambridge illustrates the passage about them with some attractive drawings but only of elementary lifting apparatus.[8] Still, neither he nor any of his contemporaries evidently regarded the invention or improvement of machinery itself as a form of applied mathematics. For at least a century before Leonardo however, notebooks had been compiled to show what was perceived as the best of contemporary designs and how new and improved versions of these machines could be devised, yet again without any explicit mathematical allusions. In Leonardo's native Tuscany a tradition of machine books and inventions goes back to the early years of the fifteenth century and is associated with the names of Fontana, Taccola, and later Francesco di Giorgio Martini.[9] But their writings are often thin on larger philosophical explanations, and again, they are silent on the mathematical–mechanics connection.

If an artist-engineer in this tradition, such as Leonardo, could think of machines as composed of elements that could be defined and explained mathematically, it was thought that mathematicians themselves might be just as qualified to take up the field. Machines exemplified simple geometrical rules, and therefore it should have been possible for those who understand those rules to improve the performance of the machines they saw in everyday use, or indeed invent quite novel machines. But there is a certain reflexive law apparent in the thinking of at least the late sixteenth century, which may help explain Leonardo's remark nearly a century earlier.

In effect, before Leonardo there were two traditions: one involving mathematical techniques and the use of instruments for measurement and a separate, flourishing art of constructing mills, cranes, and other machines. There was a link between the two, however, in the profession of architecture, since the erection of large buildings required at the least the skill to make and use machines for raising or hauling weights. Pumps might also count, because they raise liquid weights and could be utilized to supply water to buildings. That is why the most influential of ancient authors on technological matters, the architect Marcus Pollio Vitruvius, had devoted an entire book to machines. These were the machines that the architect should know and know how they should be constructed and employed. He opens Book X with engines used directly in moving and raising building materials such as

[8]Leon Battista Alberti, *De Re Aedificatoria* (1452; first printed Florence: Nicolai Laurentii, 1485), VI.6–8; translated as, *On the Art of Building in Ten Books* (Cambridge, Mass: MIT Press, 1988).

[9]For Taccola, see Frank D. Prager and Gustina Scaglia, *Mariano Taccola and his Book* de Ingeneis (Cambridge Mass.: MIT Press, 1972) and Mariano Taccola, *De Rebus Militaribus (De machinis, 1449)*, ed. Eberhard Knobloch, Saecula Spiritalia Ingenieria e Arte Militare 11 (Baden-Baden: V. Koerner, 1984). For Francesco di Giorgio see C. Maltese (ed.), *Francesco di Giorgio Martini, Trattati di Architettura, Ingenieria e Arte Militare*, Trattati di architettura 3 (Milan: Il Polifilo 1967), and F.P. Fiore, *Cittá e Macchine del '400 nei Disegni di Francesco di Giorgio Martini*, Studi of the Accademia toscana di scienze e lettere La Colombaria 49 (Florence: L.S. Olschki, 1978). And for all these artist-engineers, see Pamela O. Long, *Openness, Secrecy, Authorship: Technical Arts and the Culture of Knowledge from Antiquity to the Renaissance* (Baltimore: Johns Hopkins University Press, 2001), esp. ch 4ff.

sheer-legs (a type of A-frame), hoists, and winches. He also then describes machines use to raise water: the screw-pump (*i.e.*, the "Archimedean screw", though not called Archimedean by Vitruvius, incidentally); a waterwheel, which he termed a *tympanum*; a watermill, in which the wheel is turned by the flow of water and the power transferred through gearing to drive a millstone; and a force-pump. Vitruvius' reference might even have given the humble force-pump, like the screw-pump, an aura of classical antiquity.

Originally Vitruvius' waterwheel was intended to raise water from a river for irrigation or to supply a town's aqueducts. If the current was strong enough, the stream might suffice to turn the wheel for drainage, a device which has survived to the present day in some places, and usually known under its Arabic name, *noria*. More commonly, these wheels had to be moved by animal muscle power, where still water needed to be either drained from a pond or raised to irrigate a field. In effect, a mill employs a waterwheel in reverse, for now the flow of water is made to perform work, at first for grinding grain, as Vitruvius explains. Much as authors of Renaissance treatises on architecture like Alberti, Serlio, and Philibert de l'Orme might have admired Vitruvius, they do tend to concentrate rather on recovering his aesthetics in the outward appearance of their buildings. Understandably, they were concerned to impress patrons with their knowledge of how to produce a harmonious and elegant façade. They therefore treated machinery as of lesser importance, even though pumps and cranes had actually been improved in their recent times. Indeed, waterwheels were then employed more widely than in Vitruvius' time and in a far more diverse range of industries. Even so, that tenth book of Vitruvius remained as a guide and a possible reference that could encourage modern builders to innovate.

Despite Vitruvius and the legend of Archimedes, the practical work of mechanics, and the mathematical investigation of scholars remained effectively unlinked through the first third of the sixteenth century. Then, in the middle years of that century, three topics burst on the intellectual scene, which in association might have opened the way to this discourse. Of these the most significant was the publication by Niccolò Tartaglia of his *Nova Scientia* in 1537. Whatever later generations might think of the discoveries he claims there, this revealed to a wide public the immense potential of the mathematical approach to dynamic problems. In his *Quesiti et Inventioni* of 1546, he speaks highly of the *Mechanical Problems*, as "very good and certainly most subtle and profound in learning", but nevertheless feels that some of Aristotle's assertions can be criticized, at least in the light of the medieval science of weights.[10] This particular passage, like others in the *Quesiti*, is in the form of a dialogue, in which Tartaglia answers queries by a collocutor, in this case an eminent Spanish nobleman and diplomat, Diego Hurtado de Mendoza, who had acquired a

[10]Niccolo Tartaglia, *Quesiti e Inventioni diverse* (Venice: Lulio, 1546). The seventh book is devoted to a discussion of the "questioni mechanice, which he avers is "benissimo et certamente le sono cose suttilissime et di profonda dottrine." See also Matteo Valleriani, Lindy Divarci, and Anna Siebold, *Metallurgy, Ballistics, and Epistemic Instruments: the* Nova scientia *of Nicolò Tartaglia* (Berlin: Edition Open Access, 2013) and Raffaele Pisano and Danilo Capecchi, *Tartaglia's Science of Weights and Mechanics in the Sixteenth Century*, History of Mechanism and Machine Science 28 (Dordrecht: Springer, 2015).

copy of the *Mechanical Problems* in Greek that he was translating into Spanish.[11] De Mendoza and Tartaglia exemplify a revived interest in mechanics as a mathematical science, and they and other like-minded individuals about mid-century are likely to have inspired, or at least encouraged Alessandro Piccolomini—more a humanist than a mathematician—to produce a free translation of the *Mechanical Problems* into Latin.[12] He quite honestly called it a "paraphrase" to which he added his own exposition.

Others soon took up the task of developing mechanics in a clearer fashion, with more logical geometric treatments, such as Giovanni Battista Benedetti and Guid'Ubaldo del Monte. Guid'Ubaldo was more original in that he tried to re-establish what we would term statics on a firmer basis, structured as a theoretical interpretation of the simplest machines.[13] Whereas Tartaglia had written mostly in his native Italian, these three wrote in Latin, although they evidently read Greek. They viewed the mechanic's repertoire of their day very much from the outside, although it is fair to say that they did try to show the relevance of philosophical issues to the tools and machines in common use; the books of Piccolomini and del Monte were in fact translated into Italian, with the firm implication that this material would be of value to practical men.[14]

From another direction entirely came an added element of what Leonardo so presciently saw converging at the turn of the sixteenth century. Equally practical in intention was the publication of books about mathematical instruments, primarily for survey work. Peter Apianus' *Instrument Buch*, which was written in German and clearly aimed at a wider public, appeared in 1533 and was the first general treatise on measuring instruments for astronomy and surveying.[15] In the same year, the leading instrument maker Gemma Frisius, who had previously annotated an earlier astronomical work of Apianus (the *Cosmographia*, which appeared in 1529), published what proved to be the first detailed exposition, although in Latin, of the technique of triangulation.[16] Tartaglia, too, included in *Nova Scientia* (1537)

---

[11] The manuscript was written at Trento in 1545 or 1546, but not printed. It was published by R. Foulché-Delbosch, "Mechanica de Aristoteles," *Revue Hispanique* 5 (1898): 365–405.

[12] Alessandro Piccolomini, *In Mechanicas Quaestiones Aristotelis Paraphrasis* (Rome: A. Bladum, 1547).

[13] Guido Ubaldo del Monte, *Mechanicorum Liber* (Pesaro: Hieronymum Concordiam, 1577). For his fundamental concepts, see M. van Dyck, "Gravitating toward Stability: Guidobaldo's Aristotelian-Archimedean Synthesis," *History of Science* 44 (2006): 373–407. The best introduction to this theme is still Stillman Drake and I.E. Drabkin, *Mechanics in Sixteenth Century Italy (selections from Tartaglia, Benedetti Guido Ubaldo and Galileo)* (Madison: University of Wisconsin Press, 1969).

[14] E.g., Oreste Vannocci Biringucci, *Parafrasi di monsignor Alessandro Piccolomini sopra le Mecaniche d'Aristotile* (Rome: F. Zanetti, 1582) and Filippo Pigafetta, *Le mechaniche: nelle quali si contiene la vera dottrina di tutti gli istrumenti principali da mover pesi grandissimi con picciola forza* (Venice: Francesco di Franceschi, 1581).

[15] Petrus Apianus, *Instrument Buch* (Ingolstadt: P. Apianus, 1533).

[16] Gemma Frisius' *Libellus de Locorum describendorum ratione* is attached to Apianus' *Cosmographicum liber* (Antwerp: Arnoldum Birckman, 1533).

an illustrated account of the uses of his own version of the geometric square. These works were followed by a number of books and pamphlets that all might describe and advertise the author's own ingenious and compendious invention to make precise observation and calculation as simple as possible using instruments.

Finally, one other tradition appeared in print in mid-century that extended the machine-book tradition of the fifteenth century more firmly into the humanist sphere. Although it was not primarily about machines, Georgius Agricola's classic *De Re Metallica* (1545), on mining and metallurgy, could make a third with these more straightforwardly mathematical studies, since in this lavishly illustrated treatise he covered everything that anyone of the time could conceivably have wanted to know about metals, how to extract and refine them. In doing so he describes and depicted in great detail all the machines that were used in those industries.[17] Of course, Agricola did not write for miners or foundrymen: his book was supposedly meant for serious scholars, but was really for potential investors. In writing to this audience, however Agricola shows how far mechanization had already advanced; that made it, intentionally or unintentionally, a subject in which an educated elite could take an interest. They then might even think it worthwhile to bring in such large and costly machines in the hope of saving time and money in the long run. The machines that appear in the book may be for crushing ore, for instance, but a larger proportion are for drainage or hoisting material, so they looked as if they could easily be adapted for comparable surface operation.

Whether inspired by the mathematical instrument literature, or by Agricola, or even by coming across copies of much older manuscripts—but probably not Leonardo himself—in the latter part of the sixteenth century three authors in particular took up Leonardo's proposition quite explicitly. The first person to have printed a book of mechanical inventions, Jacques Besson, was a teacher of mathematics like Tartaglia. He first appears in this role in Geneva and later in Orleans, where in 1569 he obtained a privilege for his book of inventions. He had in 1567 already published a detailed account of his very own very clever and original mathematical instrument, which he called a *cosmolabe*, though he notes that he had been planning a book of his machines for some time, all the while dabbling in theories on the origin of rivers, in pharmaceutical chemistry, and even practicing briefly as a Huguenot pastor.[18] His book proved a landmark, and was eventually

[17] Georgius Agricola, *De Re Metallica* (Basel: H. Frobenium et N. Episcopium, 1556), translated as Herbert and Lou Henry Hoover (ed. and trans.), *De Re Metallica* (New York: Dover Publications, 1950).

[18] Jacques Besson, *Le Cosmolabe ou Instrument universel concernant toutes observations qui se peuvent faire par les sciences mathématiques, tant au ciel, en la terre* (Paris: Ph. G. Deroville, 1567). In the same year he published his inventions book, he also published a more 'traditional' text on waterworks: *L'Art et science de trouver les eaux et fontaines cachées soubs terre, autrement que par les moyens vulgaires des agriculteurs & architects* (Orleans: E. Gibier, 1569). His first work was on the medicinal extracts of the olive tree: *De absoluta ratione extrahendi olea & aquas e medicamentis simplicibus* (Tiguri [Zurich]: Andream Gesnerum Jr., 1559).

published in Latin, French, Italian, German and Spanish versions.[19] His title page proclaims that this is a book of mathematical and mechanical inventions. For a start, he begins by illustrating what he explicitly, and perhaps surprisingly, names "geometrical and mechanical" instruments: a spring, a file, a screw with its nut, dividers and a ruler. This leads on to a proportional compass, complete with a dial, "for measuring the symmetry of the parts of a body," followed by other ingenious compasses, intended to convert straight to curved lines, to construct pyramidal or conical shapes, and to construct ellipses. Then he proceeds to show devices to cut oval or helical figures, as if he was thinking of very elaborate lathes. Only after that does he begin to portray his wide range of supposedly novel machines, including among them one to illustrate his own theory that would explain how Archimedes could have hauled a ship upon the shore single-handed.

So far most of the authors that have been mentioned would have seen themselves as scholars and teachers. Still, the profession that combined the skills of the architect and the application of mathematics most clearly was the military architect – or as he was coming to be known, the engineer. The missile weaponry of the Middle Ages required constructing siege towers and engines such as trebuchets and mangonels, which an *ingeniarius* might knock up from available timber, even if kings might well wish to retain a few in major fortresses, to be handy when needed. Really then, this man would have been another specialist carpenter. Once these strange engines were rendered obsolete when cannon and gunpowder displaced them, a series of newly defined and specialized engineering professions arose, which traded on the idea that what they were doing was somehow mathematical. The common founder, who owed more to the technique of casting bells then to the trade of the old *ingeniarius*, created guns using proprietary techniques. The gunners who fired the cannon soon formed another new and highly skilled trade, which gravitated towards mathematical instruments to ply their profession. Traditional fortification was rendered, if not useless, at least much more vulnerable, so the new military engineer designed fortifications to meet the threat of artillery: structures which were primarily gun emplacements, intended to expose those who manned them to hostile fire as little as might be consistent with offering a barrier to direct assault. That meant the engineer had to be a surveyor who knew how to use mathematical instruments to measure heights and distances of enemy fortifications or a besieger's encampment, which obviously he could not approach too closely.[20] Of the five authors who published books of mechanical invention in France in the latter part of the sixteenth century, four had some military experience—Besson was the only

[19]Jacques Besson, *Instrumentorum et Machinarum. Liber Primus* (Orleans or Lyons?, 1569). Only later editions are entitled *Theatrum instrumentorum et machinarum* (*e.g.*, Leiden: B. Vincentium, 1578).

[20]Christopher Duffy, *Siege Warfare; the Fortress in the Early Modern World* (London: Routledge & Kegan Paul, 1979). Horst de La Croix, "The Literature on Fortification in Renaissance Italy," *Technology and Culture* 4 (1963): 30–50 lists well over a hundred treatises specifically on fortification. He adds some books of general advice to army officers, including material on this theme, which has been the basis for continued study.

exception—and two, Agostino Ramelli and Jean Errard, were professional military engineers.[21] Interestingly, whereas the manuscript tradition had been Italian or German, as were most of the books of theoretical mechanics and technology, this development took place in France.

If the new style engineer had to have enough mathematics to handle his instruments he was also expected to know something of machines. In fact moving cannon about presented fresh problems, which encouraged commanders to make sure their artillery trains were equipped with the jacks and hoists that formed the core of contemporaneous problems in works on mechanics, themselves often furthering or deriving from the *Mechanical Problems* text itself. Jacks had actually been invented for a purpose similar to their use today, to keep wagons upright while a broken wheel was changed, but it was soon realized that they could also serve to lift a fallen or stacked gun so that a hoist could set it on its carriage. So books of advice for artillery officers point to occasions when they had come in particularly handy, while a leading writer on mechanics explains how in past wars the lack of a simple lifting device had had dangerous consequences. A contemporary English author, Thomas Smith, in his *Art of Gunnery* (1600) explains that the Master of Ordnance among other things to "have in readinesse" also requires "engines for mounting or dismounting of Ordnance, Wheeles, Axletrees....They ought also to have some sight in the Mathematicalls... to practise all Geometricall Instruments."[22] A jack could however serve as well as an offensive weapon, to lift a door or perhaps a postern off its hinges in order to break into a hostile building. It is curious but doubtless no accident that the first detailed printed illustrations of jacks appear in Ramelli and Errard, and are shown as put to that particular job. Indeed the first real depiction of such simple tools as spanners and wrenches likewise appear there and the oldest surviving examples are in armories.[23] The machines are mainly these

[21]For Ramelli, *Le Diverse et Artificiose Machine* (Paris: casa del Autore, 1588), see Eugene S. Ferguson (ed.) and Martha T. Gnudi (trans.), *The Various and Ingenious Machines of Agostino Ramelli (1588)* (Baltimore: John Hopkins University Press, 1976) and *cf.* Alex G. Keller, review of this edition as "Renaissance Theaters of Machines," *Technology and Culture* 19 (1978): 495–508. For Jean Errard, *Le Premier Livre des Instruments Mathématiques et Mécaniques* (Nancy: Jan-Janson, 1584) see Albert France-Lanord (ed.), *Le Premier livre des instruments mathématiques méchaniques* (Paris: Berger-Levrault, 1979). Both Ramelli and Errard had sons who followed them into the new profession, which suggests its attraction as a career at the time.

[22]Thomas Smith, *Art of Gunnery* (London: n.p. for William Ponsonby, 1600), 74. Gabriello Busca, *Della Architettura Militare. Primo libro* (Milan: Bordone and Locarno 1601), proposed in a 'book III' to deal with machines, explaining how necessary it was for a "military architect" to know how to make them. Diego Ufano, *Tratado de la Artilleria* (Brussels: Juan Momarte, 1612), 142 and 223–53, also insists on knowledge of machines for the gunnery officer. Pigafetta, in the dedication to the military engineer and *general d'altegliaria* of the Republic of Venice, Count Giulio Savorgnan, of his translation of Guid'Ubaldo's work (n.14, above), stresses this point. See also, A.G. Keller, "Mathematicians Mechanics and Experimental Machines in Northern Italy in the Sixteenth Century," in Maurice P. Crosland (ed.), *The Emergence of Science in Western Europe* (London: Science History Publications, 1975), 15–34.

[23]Jacks, spanners and wrenches appear in Ramelli, pl. 155–59 and the use of a jack for undermining a wall in Errard, *Le Premier Livre* (n.21, above), pl. 4.

compact lifting devices, although as an army on the march needs to keep itself supplied with food, mobile mills, which could be taken apart and reassembled, could also come in handy. The "geometricall instruments" would naturally be those in common use to measure areas, heights and distances of potential targets. Ramelli included a kind of two-way collapsible winch to be used to move cannon up a rocky mountain road. It is almost the only one of his devices that he claims to have tried out in use.[24] The literature also describes pontoon bridges and sliding bridges to cross moats and mobile mills, in which the team of horses drew the mill in a cart, together with the bars which enabled them when in camp to operate the mill. One book of inventions depicts such a device, which the author claims was employed by the Spanish army of Spinola in the Low Countries.[25]

Beside Ramelli's great collection of military and civil machines, he wrote a companion volume, all about the many uses of his own compendious instrument, his Triangle, which has survived (now at Chatsworth in England), although he was unable to publish it.[26] In the preface Ramelli begins with Plato—not without a nod to Pythagoras—claiming that his own long experience at war and his deep studies had led him to invent this new instrument, which however was in reality not so different from others devised by ingenious mathematicians of his day. Ramelli did succeed in publishing his book of various ingenious inventions, by far the biggest and most elaborate of all the sixteenth-century machine books. Even though there is little mathematical content in the main text of the book where each machine is described—not even measurements—Ramelli clearly intended his book to demonstrate the potential of a mathematics-based technology, for he included a preface in praise of mathematics, which stresses both the practical utility of these sciences over a broad range of occupations well as their reliability.

Errard too opens his book with a depiction of the law of the lever as he understood it. An inset geometrical diagram helps the reader to understand how two weights can be in equilibrium when suspended at different distances from a central point: a winch applies this ratio to the movement of a pair of spur-gears and to a worm-gear couple, with a pair of spur-gears in the same plane, meshing through teeth set radially around the rim of each wheel. He also includes measuring instruments and what seems to be meant as a universal dial and he returns to the classical *Mechanical Problems* problem of how Archimedes moved the ship, which in his solution travels on rails. In his later career he became much better known as a fortification expert,

---

[24] Ramelli, *Le Diverse et Artificiose Machine*, pl. 189.

[25] Vittorio Zonca, *Novo Teatro di Machine et Edificii* (Padua: Pietro Bertelli, 1607), btw. 88–89.

[26] Agostino Ramelli, *La Fabrica et l uso del triangolo del Capitan Agostino Ramelli dal Ponte della Tresia ingegniero del Christianissimo Re di Francia*, noted as being "on vellum with beautiful drawings"; *Catalogue of the Library at Chatsworth* (London: Chiswick Press, 1879), 4: 347. [The manuscript apparently has a printed title page, so it is possible that Ramelli tried to get the manuscript into print; *Martha Teach Gnudi Research and Publication Papers 1540–1977*, Louise M. Darling Biomedical Library History, History and Special Collections Division, University of California–Los Angeles, MS Coll. no. 307, box 2, folder 3.—ed.]

particularly on account of the practical geometry text he wrote, meant for use in his own and kindred professions.[27]

Although the books of Besson, Ramelli, Errard and others later in the seventeenth century—and perhaps Besson's in particular since he was translated into several languages—undoubtedly played a part in linking mechanical invention to the application of mathematics, they were certainly helped by leading publicists for mathematics. They included these mechanical developments among their reasons for encouraging and diffusing an understanding of mathematics. The best known and probably the most influential of these publicists was Pierre de la Ramee (known as Petrus or Peter Ramus) in his public orations in France.[28] In England we could add the names of Welshmen Robert Recorde and John Dee: Recorde in the introductions he wrote for his textbooks of arithmetic and geometry and Dee in the preface to the first English translation of Euclid's *Elements*.[29] In Britain and France alike there was a sense that they had been left behind; that the wealth of the German lands came from the more widespread knowledge of practical mathematics, particularly as related to mining and metallurgy, while the wealth and power of the Iberian peninsula came from the pursuit of mathematical navigation, and the training of pilots. The Italian states might be less powerful, although they were still exporters of talent. However transalpine Europe admired Italian achievements in the fine arts, it was fortification that provided the best living for that was what bellicose states required (indeed, which sixteenth-century states were not bellicose?). The alien technician might be hired and even admired, but he was not loved and often not really trusted. The "straungers" were suspected of keeping their knowledge to themselves instead of training locals to replace them.[30] Since the new art depended on the rationality and certainty of mathematics, the locals were convinced that they could learn as well as these foreigners. That must in part account for the huge literature on fortification, artillery and linked military skills through the sixteenth century and much of the seventeenth, promoted by the internal conflicts which tore much of Western Europe apart as a the consequence of the Protestant Reformation.

In France at least engineering apparently appealed more to Huguenots, among their number both Besson and Errard, besides Salomon de Caus and less well-known figures like Joseph Boillot. Italians might serve the Catholic side so that if one

[27] Jean Errard, *La geometrie et practique generalle d'icelle* (Paris: D. Le Clerc, 1594).

[28] *E.g.*, Petrus Ramus, *Prooemium Mathematicum* (Paris: Wechelus, 1567), 291–93. And in general, see Walter J. Ong, *Ramus, Method, and the Decay of Dialogue: From the Art of Discourse to the Art of Reason* (Chicago: University of Chicago Press, 2004) and Steven J. Reid and Emma Annette Wilson (eds.), *Ramus, Pedagogy, and the Liberal Arts: Ramism in Britain and the Wider World* (Farnham: Ashgate, 2011).

[29] Robert Recorde, *The Grounde of Artes, Teachyng the Worke and Practice of Arithmetike* (London: Reginalde Wolfe, 1551) and *The Pathway to Knowledge, Containing the First Principles of Geometrie* (London: Wolfe 1551). John Dee, preface to *Euclid, The Elements of Geometrie*, trans. Henry Billingsley (London: Iohn Daye, 1570), for which see, *The mathematicall praeface to the Elements of geometrie of Euclid of Megara (1570)*, intro. Allen G. Debus (New York: Science History Publications, 1975).

[30] Henry Heller, *Anti-Italianism in Sixteenth Century France* (Toronto: University of Toronto Press, 2003).

Italian, Ramelli, served the French royal cause at the first great siege of La Rochelle, another, Pompeo Targone, turns up at the second. Italians were even prepared to cross the religious divide, like Genebelli who worked first at Antwerp, when the city was in rebellion against the Spanish crown, and then moved to England. Jacopo Acontio might be regarded in a similar light, although in his case he realized that this would be the best profession to follow because he had become a Protestant at heart and knew that he would have to seek employment in Protestant lands. Both Genebelli and Acontio felt they could work equally in civil projects as engineers and sought reward for their original mechanical inventions, which for example could be used in drainage and dredging operations. Evidently they believed they could turn their skill to this kind of work and to mechanical improvement as well as to military tasks.[31]

In fact quite a number of mathematical writers announced that they had devised inventions, mostly mechanical, which they hoped to publish, so that the number of inventors on paper is greater than the number who actually published and some notable collections were never printed. Probably the sheer cost of publication made it very difficult to bring out books that depended on lavish and precise illustration unless the author could obtain a generous and preferably royal subsidy. Among those who promised to give the world their machine books we may name Oreste Biringucci, in his translation of Piccolomini's paraphrase of the *Mechanical Problems*.[32]

It is true that neither textbook writers nor machine book authors were necessarily practical men. Besson does not appear to have tried to construct any of his inventions.[33] Expensive as his book must have been to produce, it would be much easier to do that and even to make some models, as he appears to have done, than to build large machines (not least because although the scale effect was known, it was not properly understood). In many ways it was easier for those with *less* practical experience to propose machines as mathematical devices.

There is a curious paradox here: rulers, patrons, and investors could be persuaded that geometrical analysis would lead to mechanical improvement before the ordinary machine makers would accept such a notion. The mathematization of machine design did not happen overnight, nor as universally as mathematical techniques became essential for the survey of land or the ground plans of buildings. Indeed,

---

[31]See A.G. Keller, "Aconcio, Jacopo (*c.*1520–1566/7?)," in *Oxford Dictionary of National Biography*, eds. H. C. G. Matthew and Brian Harrison (Oxford: OUP, 2004); online ed., ed. Lawrence Goldman, January 2014, and Lynn White, Jr., "Jacopo Acontio as an Engineer," *American Historical Review* 72.2 (1967): 425–444. Although Genebelli appears in nearly all accounts of the siege of Antwerp, and after his move to England in various official records, there does not seem to be any survey of his curious career. For some notice of his activities, see Steven A. Walton, "State Building through Building for the State: Domestic and Foreign Expertise in Tudor Fortifications," in Eric Ash (ed.), *Expertise and the Early Modern State*, Osiris 25 (2010): 66–84.

[32]Biringucci, *Parafrasi ... Piccolomini* (n.14, above), 6.

[33]A.G. Keller, "The Missing Years of Jacques Besson, Inventor of Machines, Teacher of Mathematics, Distiller of Oils and Huguenot Pastor," *Technology and Culture* 14 (1973): 28–39.

one could safely say that the concept only really came to fruition in the course of the eighteenth century or even later. Nevertheless, the idea, at least in the sense of that potentiality, was there from a much earlier time and was promoted as much by this strange literary genre of the artist-engineers as by the scholars who urged a greater role for mathematical learning. Both hoped and probably needed royal or at least noble support. Naïve, limited, and eccentric as they may seem to us, those who first treated machines as *improvable* mathematical devices launched a project which was to have great bearing on the future development of engineering and industry. Indeed the very term 'mechanical philosophy' implies that this approach provided a model for describing natural processes as well as man-made inventions.

Part III
What Was the Relationship Between Practical Mathematics and Natural Philosophy?

Chapter 7
The Making of Practical Optics: Mathematical Practitioners' Appropriation of Optical Knowledge Between Theory and Practice

Sven Dupré

Abstract The discussion of the differing practices of mathematical practitioners' appropriation of the optical tradition in this essay brings out a variety among mathematical practitioners and *within* the tradition of practical mathematics. This diversity is difficult to grasp in accounts of practical mathematics which oppose theory and practice as mutually exclusive categories. Comparing the optical projects of two geographically and socially differentiated mathematicians, the Venetian physician and mathematician Ettore Ausonio and the English town councilman and volunteer gunner, William Bourne, this essay argues that mathematical practitioners' appropriation of optical knowledge depended upon the complexities of personal and local contexts, such as the perception of patronage opportunities. Notwithstanding the cognitive similarities of their optical projects, the balance of theory and practice is different in the presentation of their shared knowledge. Ausonio's practical optics, which aimed at the design of an instrument by offering a theoric, is contrasted with Bourne's project for the making of a telescope, which lacked any attempt at a theoric. The essay shows that, rather than as an established category, practical optics should be understood as the result of a construction by Renaissance mathematical practitioners' appropriations of the perspectivist optical tradition.

7.1 Introduction

What, if anything, do a Venetian university-educated mathematician, alchemist and physician and an English jurat, or town councillor, from Gravesend on the lower Thames, both active in the 1560s and 1570s, have in common? If we are allowed to refer to the English jurat as a 'mathematical practitioner', would it make sense to refer to both men as 'mathematical practitioners' and to the kind of

S. Dupré (✉)
Utrecht University, Utrecht, The Netherlands
e-mail: s.g.m.dupre@uu.nl

L.B. Cormack et al. (eds.), *Mathematical Practitioners and the Transformation of Natural Knowledge in Early Modern Europe*, Studies in History and Philosophy of Science 45, DOI 10.1007/978-3-319-49430-2_7

knowledge that they produced and advertised as 'practical mathematics'? If so, what do 'mathematical practitioner' and 'practical mathematics' mean in these contexts? It is with these types of questions that this essay is concerned. I will argue that the identity of 'mathematical practitioners' and 'practical mathematics' depended upon the complexities of personal and local contexts.

It is worth making this point, because recently the notion of 'mathematical practitioner'—which originated in the 1950s with Eva Taylor's work on Tudor and early Stuart England[1]—has been critically re-examined. Eric Ash has argued that the idea of the emergence of a new intellectual and cultural community of 'mathematical practitioners'—one in which practical mathematical knowledge as well as a common perception of the aims of that discipline were shared—obscures both the diversity of social and intellectual identities of these practitioners and the interactions and rivalries *within* this group.[2] Ash's point is that the mathematization of practical arts in early modern England helped to divide rather than to form a community. We will have the occasion to point to similar processes of differentiation between proponents of mathematization of practical arts and 'practitioners' in this essay. Adam Mosley has added a comparative dimension to this point with his argument that outside the urban and entrepreneurial context of Elizabethan London, inhabited by the English 'mathematical practitioner', it was not unusual for instruments to function as models instead of only as calculating devices.[3] This suggests that there is something specifically 'English' about the 'mathematical practitioner'.

While I share Mosley's concern for the use and meaning of 'mathematical practitioner' as an actors' category, I also think that, for analytical purposes, it is still useful to speak of a field of 'practical mathematics' across contexts, and even 'national' boundaries. This notion of 'practical mathematics' is sustained by shared knowledge in the form of common instruments, tools, approaches and concepts, of which we will find examples in this essay. It is then possible—as I will do by way of the example of practical optical knowledge—to use the notion of 'appropriation' to get at the differences and varieties *within* the tradition of practical mathematics. In this essay, I will thus argue that the discussion of the differing practices of mathematical practitioners' appropriation of the optical tradition brings out this variety among mathematical practitioners and *within* the tradition of practical mathematics. I will show that mathematical practitioners' appropriation of optical knowledge depended upon the complexities of personal and local contexts, such as the perception of patronage opportunities.

[1]E. G. R. Taylor, *The mathematical practitioners of Tudor and Stuart England* (Cambridge, 1954).

[2]Eric H. Ash, *Power, Knowledge, and Expertise in Elizabethan England* (Baltimore and London, 2004), 140–142.

[3]Adam Mosley, "Objects of knowledge: Mathematics and Models in Sixteenth-Century Cosmology and Astronomy," in *Transmitting Knowledge: Words, Images and Instruments in Early Modern Europe*, eds. Sachiko Kusukawa and Ian Maclean, 41–71 (Oxford, 2006).

Elsewhere I have used the notion of 'appropriation' to argue that the image of optics changed as a consequence of Renaissance mathematicians' appropriation of optical knowledge.[4] It is well-known that the fundamental aim of perspectivist optics was the study of vision, not only of visual perception, but also of visual cognition.[5] The image of optics which Renaissance mathematicians endorsed was one that exchanged the medieval concerns about vision, perception and cognition for a focus on the design of optical objects such as mirrors. If, and how, Renaissance mathematicians' appropriation of medieval optics was influential for the development of 'geometrical optics' in the seventeenth century—in accordance with a larger interpretative framework which has attributed a privileged role to 'mathematical practitioners' and practical mathematics in the transformation of science and the making of the 'scientific revolution' in other domains of natural philosophical knowledge[6]—is still an open question.

John Schuster's work on Descartes' optics has suggested that this transformation should also be understood in terms of an *appropriation* or a physico-mathematical reading of practical mathematics.[7] The notion of appropriation might serve not only to elucidate the modes of causation connecting practical mathematics and natural philosophy, however, but also to show the diversity and plurality *within* the tradition of practical mathematics. Roger Chartier has convincingly argued, in this connection, for placing the notion of appropriation "at the centre of a cultural historical approach that focuses on differentiated practices and contrasted uses".[8]

The sort of variety within the tradition of practical mathematics with which I will be concerned here is the diversity of balances between theory and practice. This diversity has recently been recognized in different practical mathematical contexts: Alison Sandman and Eric Ash have argued that in transferring his expertise from Spain to England Sebastian Cabot created a new balance between theory and practice in navigation,[9] while Stephen Johnston has criticized the portrayal in the literature of a homogeneous English tradition of magnetism prior to William Gilbert's 'theoretical' framework expanding on the more limited 'practical' tradition

[4]Sven Dupré, "Ausonio's Mirrors and Galileo's Lenses: The Telescope and Sixteenth-century Practical Optical Knowledge," *Galilaeana: Journal of Galilean Studies*, 2 (2005): 145–180.

[5]A. Mark Smith, "Getting the Big Picture in Perspectivist Optics," *Isis*, 72, (1981): 568–589.

[6]J. A. Bennett, "The Challenge of Practical Mathematics," in *Science, Culture and Popular Belief in Renaissance Europe*, eds. S. Pumfrey, P. L. Rossi and M. Slawinski, 176–190 (Manchester, 1991); J. Bennett, "Practical Geometry and Operative Knowledge," *Configurations*, 6, (1998): 195–222.

[7]John A. Schuster, "Descartes *Opticien*: The Construction of the Law of Refraction and the Manufacture of its Physical Rationales, 1618–1629," in *Descartes' Natural Philosophy*, eds. Stephen Gaukroger, John Schuster and John Sutton, 258–312 (London and New York, 2000); and also his contribution "Consuming and Appropriating Practical Mathematics" to this volume.

[8]Roger Chartier, *Cultural History: Between Practices and Representation* (Cambridge, 1988), 13.

[9]Alison Sandman and Eric H. Ash, "Trading Expertise: Sebastian Cabot between Spain and England," *Renaissance Quarterly*, 57 (2004): 813–846.

of the mathematical practitioners.[10] Johnston has questioned, on the one hand, the usefulness of categories of 'theory' and 'practice'—a 'theoric' is at the same time both a geometrical and 'theoretical' model and a 'practical' instrument for predictive calculation, for example—and, on the other, the homogeneity of the English tradition of practical mathematics. In this essay I will discuss the diversity of balances between practice and theory within the tradition of a branch of practical mathematics, *practical optics*.

This diversity appears in the comparison of the optical projects of two geographically and socially differentiated mathematicians, the Venetian physician and mathematician Ettore Ausonio (ca. 1520 – ca. 1570) and the English town councilman and volunteer gunner, William Bourne (ca. 1535–1582). Not only did the two men never meet, they almost certainly had no notion of each other's work. Thus, we can hardly consider them members of the same community, and certainly they would not have identified themselves as such. Nevertheless, their optical projects show such similarities in scope and aims as to be recognized as practical mathematics. One of the most striking similarities is their focus upon (and limitation to) the design of an 'instrument,'—respectively a concave mirror and a telescope—for endeavours of this sort are often considered central to the fashioning of a mathematical practitioner's identity.[11] What makes Ausonio and Bourne attractive choices for a comparative approach is their *shared knowledge* of image formation in mirrors and lenses, as I will show in the next section. However, notwithstanding the cognitive similarities of their optical projects, the balance of theory and practice is different in the presentation of their shared knowledge. I will show that these differences are best construed as variant 'readings' of the optical tradition.

7.2 Shared Optical Knowledge

What knowledge did Ausonio and Bourne share on the issue of image formation in concave mirrors and convex lenses? In the 'Theorica speculi concavi sphaerici' (ca. 1560) (Fig. 7.1) Ausonio noted that the point of combustion (at ¼ diameter of the concave mirror, where the paraxial parallel rays come together after reflection

[10]Stephen Johnston, "Theory, Theoric, Practice: Mathematics and Magnetism in Elizabethan England," *Journal de la Renaissance*, 2 (2004): 53–62. The heterogeneity of the English mathematical tradition is also suggested in Stephen Johnston, "Like Father, Like Son? John Dee, Thomas Digges and the Identity of the Mathematician," in *John Dee: Interdisciplinary Studies*, ed. Stephen Clucas, (Springer, forthcoming).

[11]Stephen Johnston, "Mathematical Practitioners and Instruments in Elizabethan England," *Annals of Science*, 48 (1991): 319–344. However, the role of instruments in achieving mathematical authority was contested in the seventeenth century. See Katherine Hill, "'Juglers or Schollers?': Negotiating the Role of a Mathematical Practitioner," *British Journal for the History of Science*, 31(1998): 253–274.

Fig. 7.1 Magini's edition of *Ausonio's 'Theorica'*. Giovanni Antonio Magini, *Theorica Speculi Concavi Sphaerici*, (Bononiae: Apud Ioannem Baptistam Bellagambam, 1602, shelfmark 11. Fisica Cart. IV. n. 64) (By permission of the Biblioteca Comunale dell' Archiginnasio, Bologna)

in the mirror) is the locus of the 'point of inversion' (where the orientation of the image changes).[12] When the eye is closer to the mirror, the image appears upright; when it is farther away from the mirror than the 'point of inversion,' the image is inverted. Ausonio added that the point of combustion is also the locus of a maximally magnified (and maximally confused) image. When the eye is placed near this point, the perceived image is at its largest. But when the eye is then farther removed from the concave mirror than this point, the perceived image 'explodes.' Although less important here, it is interesting to note that this knowledge was not present in the medieval optical tradition. In perspectivist optics, the locus of this point was *either* a point of inversion *or* a point of combustion, but it could never be both at the same time. Witelo, for example, located the point of combustion at a different locus.[13]

In a letter written to Lord Burghley around 1580, Bourne discussed the same properties of concave mirrors and convex lenses, but he employed a different terminology, referring to 'burning beams' and 'perspective beams'.[14] In his description of concave mirrors, he noted that the image fills the whole surface of the concave mirror when the eye is placed at a certain distance from the mirror.

> And then this glasse, the property of yt ys, to make all thinges which are seen in yt to seem muche bigger then yt ys to the syghte of the Eye, and at some appoynted distance, from the glasse, accordinge to the forme of the hollowness, the thinge will seem at the biggest, and so yow standinge nearer the thinge will seeme less, unto the sighte of the eye: so that, accordinge unto the forme of the concavity or hollowness, and at some appointed distance from hym that looketh into the glasse, And yf that the glasse were a yearde broade, the beame that shoulde come unto his eye, shall showe his face as broade, as the whole Glasse.[15]

While the locus of this point, where the image fills the whole surface of the mirror, seems not to be well defined in Bourne's description of the concave mirror, as he implicitly stated that this point is close to the point of inversion of the mirror, his location of this point is much more precise in the case of convex lenses. The locus of the point of combustion is determined by the 'burninge beame,' and Bourne identified the locus of the point of combustion with the locus of the point of inversion in a description of the optical properties of the 'glass'.

> And yf that yow doo beholde any thinge thorowe this Glasse, and sette the glasse furder from yowe then the burning beame, and so extendinge after that what distance that yow list,

---

[12]I have used this version: Ettore Ausonio, "Theorica speculi concavi sphaerici,", in Galileo Galilei, *Le Opere di Galileo Galilei* (Firenze: G. Barberà Editore, 1968), 3:865. For a more detailed optical analysis of Ausonio's *Theorica* and its place in the optical tradition, see Dupré, "Ausonio's Mirrors and Galileo's Lenses", 160–170.

[13]*Ibid.*, 159.

[14]Bourne to Burghley, ca. 1580, in British Library (London), MS Lansdowne 121, item 13, ff. 96–102. Citations are taken from the edition in Albert Van Helden, *The Invention of the Telescope* (Philadelphia, 1977), who published the text from J. O. Halliwell, *Rara mathematica* (London, 1839).

[15]*Ibid.*, 32.

> all suche thinges, that yow doo see or beholde, thoroughe the glasse, the toppes ys turned downwardes.[16]

Bourne differentiated this 'burninge beame' from the 'perspective beam,' and specified the location of this 'perspective beame' *vis-à-vis* the point of combustion.

> The quality of this Glass, ys, if that the sunne beames do pearce through yt, at a certayne quantity of distance, and that yt will burne any thinge, that ys apte for to take fyer: And this burnynge beame, ys somewhat furder from the glasse, then the perspective beame.[17]

The 'perspective beame' locates the point where the eye is to be placed in order to perceive the largest possible image, one that fills the entire diameter of the lens before the image completely collapses when the eye is then placed at the point of combustion of the convex lens.[18]

> The quality of the Glasse, (that ys made as before ys rehearsed) ys, that in the beholding any thinge thorowe the glasse, yow standinge neare unto the Glasse, yt will seeme thorow the glasse to bee but little bigger, then the proportions ys of yt: But as yow do stande further, and further from yt, so shall the perspective beame, that commeth through ye glasse, make the thinge to seeme bigger and bigger, untill such tyme, that the thinge shall seeme of a marvellous bignes: Whereby that these sortes of glasses shall much proffet them, that desyer to beholde those things that ys of great distance from them. ... And allso standing further from the glasse yow shall discerne nothing thorowe the glasse: But like a myst, or water: And at that distance ys the burninge beame, when that yow do holde yt so that the sunne beames doth pearce thorowe yt. And allso yf that yow do stande further from the glasse, and beholde any thinge thorowe the glasse, Then you shall see yt reversed and turned the contrary way, as before ys declared.[19]

It is clear that Bourne discussed the same properties of the convex lens that Ausonio described for the concave mirror, albeit in a different terminology. Both mathematicians presented their discussions of the locus of image formation within the context of the advertisement of an 'instrument' of sorts. Ausonio's 'Theorica' was a one-page folio, a text and an image presumably intended as a manual to accompany the concave mirrors themselves which Ausonio had designed and upon which his contemporary reputation primarily rested.[20] When around 1560 Ausonio delivered concave mirrors to the Duke of Savoy, Emmanuele Filiberto, it is most likely that they were presented to the Duke together with a manuscript copy of

---

[16]Bourne to Burghley, ca. 1580, in Van Helden, *Invention*, 33.

[17]*Ibid*., 33.

[18]It should be noted that the loci of the 'perspective beam' and the 'burning beam' are only approximately correct. In fact, the point where the eye is to be placed to perceive the largest possible image is a very short distance beyond the point of combustion of the convex lens. Bourne was not able to locate the point of combustion more precisely.

[19]*Ibid*., 33.

[20]'Etor Eusonio da Venetia inventore delle piu belle materie matematiche che mai si sieno viste ne udite al mondo: percioche ha fatto certi specchi concavi di estimabile grandezza, ne i quali se veggono cose maravigliose'. Leonardo Fioravanti, *Dello specchio di scientia universale* (Venetia, 1678), 55v.

the 'Theorica'.[21] Bourne's letter was likewise designed to persuade Burghley of the feasibility of a telescope design. Given the magnifying properties of concave mirrors and convex lenses, Bourne was convinced that the effect would be additive if a concave mirror and a convex lens were combined.

> And so reseaved from one glasse into another, beeyinge so placed at such a distance, that every glasse dothe make his largest beame. And so yt ys possible that yt may bee helpped and furdered the one glass with the other, as the concave lookinge glasse with the other grounde and polysshed glasse. That yt ys likely yt ys true to see a small thinge, of very greate distance.[22]

Bourne's knowledge of the imaging properties of concave mirrors and convex lenses informed his telescope design.[23] (See Fig. 7.2) He was first of all explicit about the distance between the convex lens and the concave mirror, arguing that they should be placed so 'that every glasse doth make his largest beame'. Since the image of a single optical component (a concave mirror or a convex lens) was largest at the locus of the 'perspective beame', and since this locus was near the

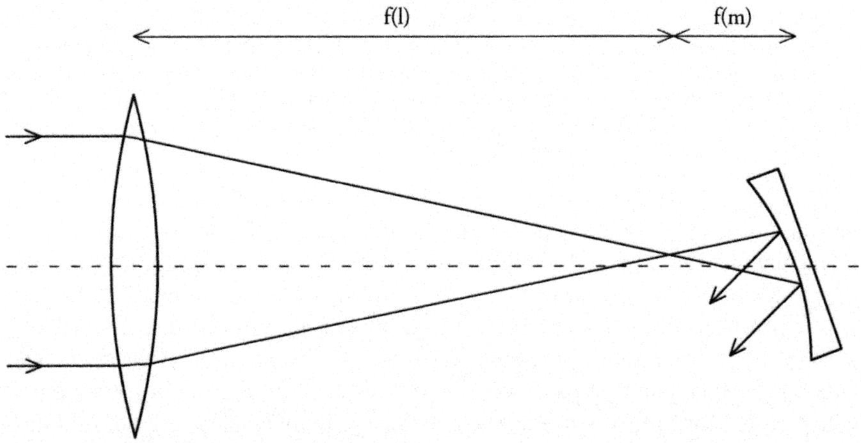

Fig. 7.2 William Bourne's telescope design, ca. 1580. (f(l) = focal length of the convex lens; f(m) = focal length of the concave mirror)

[21]R105 Sup. (Biblioteca Ambrosiana, Milan), fols. 292r–292v.

[22]*Ibid.*, 34.

[23]The design proposed here is identical to the one suggested by Colin A. Ronan, "There Was an Elizabethan Telescope," *Bulletin of the Scientific Instrument Society*, 37 (1993): 2–3; Joachim Rienitz, "'Make Glasses to See the Moon Large': An Attempt to Outline the Early History of the Telescope," *Bulletin of the Scientific Instrument Society*, 37(1993): 7–9; Ewen A. Whitaker, "The Digges-Bourne telescope – An Alternative Possibility," *Journal of the British Astronomical Association*, 103 (1993): 310–312. However, I do not share these authors' (especially, Ronan's) conclusion that this obliges us to attribute the invention of the telescope to the 'Elizabethan English'. My point (which went unnoticed in the work of Ronan) is that practical optical knowledge is at the basis of the design.

locus of the point of combustion, the distance between the convex lens and the concave mirror was determined by the focal planes of the mirror and the lens. Thus, in Bourne's telescope design the total length of the instrument equalled the sum of the focal lengths of the mirror and the lens. Secondly, Bourne's optical knowledge about the locus of a maximally magnified image is the basis of his selection of a large diameter lens. As we have seen, Bourne knew that the magnified image fills the complete surface of the lens when the eye is placed at the point of combustion or point of inversion. It is then reasonable to consider magnification dependent upon the diameter of the lens, instead of upon its focal length, and to search out the largest possible diameter. "The broader the better" was indeed Bourne's advice for the diameter of the lens.[24]

Bourne's and Ausonio's optical projects share the instrumental focus, but also the content and the type of knowledge that they regarded as pertinent to the design of mirrors and telescopes. This kind of knowledge is *practical* knowledge.[25] On the one hand, it is different from the theoretical knowledge embodied in the perspectivist tradition because it is based on familiarity with the behaviour of *real* objects, as for example, the perception of images in concave mirrors and convex lenses. On the other, it is knowledge on paper. Notwithstanding the misleading nineteenth-century title attributed to Bourne's letter—which speaks of "glasses for optical purposes, according to the making, polishing, and grinding of them"—the practical knowledge in Bourne's and Ausonio's optical works is to be differentiated from *material* knowledge. Unlike practical knowledge, material knowledge is information about how to make an instrument or how to translate the drawing on paper into a physical object. A typical example would be, in the case of mirror making, information about the kind of glass to be used.

Granted that Bourne and Ausonio share practical knowledge of image formation in concave mirrors and convex lenses, the context of presentation of this knowledge is substantially different in its appeal to theory and to the optical tradition. As we will see, this appeal is considerably more present in Ausonio's 'Theorica'. It would however be misleading to characterize the differences between the optical tradition and Ausonio, and between Ausonio and Bourne, in crude terms of theory

---

[24]William Bourne, *Inventions or Devices: Very Necessary for All Generalles and Captaines, or Leaders of Men, as well by Sea as by Land* (London, 1578), 96. Bourne insisted that the lens 'must bee made very large, of a foote, or 14. to 16. inches broade', thus of 30 to 40 centimetres, a requirement clearly beyond the contemporary technological capacities. Bourne's knowledge was practical, not material (see below).

[25]For this concept of practical knowledge (and the distinction with material knowledge), compare the discussion of practical knowledge of trajectories of cannon balls in Jochen Büttner, Peter Damerow, Jürgen Renn and Matthias Schemmel, "The challenging images of artillery: Practical knowledge at the roots of the scientific revolution," in *The Power of Images in Early Modern Science* (Basel, Boston, Berlin, 2003), eds. Wolfgang Lefèvre, Jürgen Renn and Urs Schoepflin, 3–27. For these two types of knowledge, see also Catherine Eagleton, "Medieval Sundials and Manuscript Sources: The Transmission of Information about the Navicula and the Organum Ptolemei in Fifteenth-Century Europe," in *Transmitting Knowledge: Words, Images and Instruments in Early Modern Europe*, eds. Sachiko Kusukawa and Ian Maclean, 41–71 (Oxford, 2006).

and practice. In the next section, I will consider Ausonio's work as an *appropriation* of the perspectivist tradition and Bourne's project as a response to a *reading* of the optical tradition. Insofar as possible, I will connect the different results of Ausonio's and Bourne's appropriation of the optical tradition to the differences in the contexts to which they responded.

7.3 William Bourne versus Ettore Ausonio: Theory and Practice

Ausonio was university-educated—he studied medicine at the University of Padua in the 1540s—and a practising physician in Venice who taught mathematics privately.[26] He was involved in the workshops at the Fondaco dei Tedeschi in Venice, in the printing house of Michele Tramezzino, and—as Regent of the *Stanza delle Matematiche*—in the Venetian *Accademia della Fama*. One of the mathematical topics that Ausonio taught privately was optics. His teaching was based above all on his reading of Witelo's *Perspectiva*, but his lecture notes reveal highly selective reading practices.[27] These notes listed only Witelo's descriptions of the instruments to measure reflection and refraction and those propositions in which the Polish perspectivist claimed the use of these instruments as proof of the proposition. Ausonio left out all of Witelo's propositions not established with the instruments, and in the selected propositions, he discarded the geometrical demonstrations and the geometrical diagrams. In sum, Ausonio appropriated Witelo's optics in such a way that optics appeared to be a mathematical art based on instrumental proof.

Ausonio's optics was based on an intimate though selective engagement with the optical tradition. As we have seen in the previous section, Ausonio's work contains practical knowledge of image formation in concave mirrors: his 'Theorica' is thus not to be considered theory. It is, nevertheless, a selective engagement with the optical tradition. Rather than theory, however, the 'Theorica' is a geometric *theoric* which shows how to manipulate a concave mirror to obtain mirror images. It shares several characteristics with the well-established *theorica (planetarum)* tradition of astronomy and mathematics, which makes it different from the medieval optical tradition.[28] First, Ausonio took from the medieval optical tradition the

[26]For Ausonio's biography, see Sven Dupré, "The Dioptrics of Refractive dials in the Sixteenth Century," *Nuncius*, 18 (2003): 53–57.

[27]*Ibid.*, 58–60.

[28]On the theoric, see Olaf Pedersen, "The Decline and Fall of the Theorica Planetarum: Renaissance Astronomy and the Art of Printing," *Studia Copernicana*, 16 (1978): 157–85; Jim Bennett, "Knowing and Doing in the Sixteenth Century: What Were Instruments For?," *British Journal for the History of Science*, 36 (2003): 142–43. On the visualizations in the *theorica planetarum*, see Isabelle Pantin, "L' illustration des livres d' astronomie à la Renaissance: L' évolution d' une discipline à travers des images", in *Immagini per conoscere: Dal Rinascimento al Rivoluzione Scientifica*, eds. Fabrizio Meroi and Claudio Pogliano, 3–41 (Firenze, 2001).

geometry of image formation—known as the cathetus rule—but unlike opticians within the medieval tradition, and like those within the theoric tradition, he was not interested in demonstrations. Secondly, and again unlike medieval students of optics, Ausonio did not use geometrical diagrams. His drawing suggests the three-dimensionality of a real mirror. It is this suggestion of physicality that Ausonio's 'Theorica' shares with other theorics. These visual characteristics also made his work attractive to courtly patrons. In the 1560s the Duke of Savoy, Emmanuele Filiberto, asked Ausonio to equip his library with a collection of mathematical and optical instruments, including a concave spherical mirror. It is likely that Ausonio's 'Theorica' functioned as a kind of manual accompanying the real concave mirror travelling to the library of Filiberto. The optical knowledge in the 'Theorica' was then packaged so as to respond to the demands of an instrument collection of a princely library and to appeal to the patronage of the Duke of Savoy.[29]

In contrast to Ausonio, Bourne did not attempt to establish a theoric of image formation. This resulted in several differences from Ausonio's theoric. Not only is Bourne's optics lacking all demonstrations—like Ausonio's theoric—Bourne also dismissed all references to the medieval optical tradition. Unlike Ausonio, Bourne did not cite the work of Witelo—it is doubtful that he even knew perspectivist optics—nor did he make use of any of the geometrical models and rules, such as the cathetus rule, which perspectivist optics had developed for representing and predicting the locus of images. Although he did not give geometrical demonstrations, Ausonio referred to the propositions in Witelo's optics underlying his theoric of image formation in a concave mirror. Moreover, Ausonio used the terminology of perspectivist optics—lines of incidence, cathetus of incidence, lines of reflection, and even that of a 'physics' of light and vision, *res forma, species intentionales*—although the latter did not play any role in the making of his theoric.[30] Bourne, in contrast, made no references to a 'physics' of light and his terminology of "burning beam" and "perspective beam" did not derive from the optical tradition. Finally, Ausonio's image of a concave mirror, which could be used as a practical instrument to facilitate the prediction of the locus of an image given the locus of the eye and the object, had no equivalent in Bourne's letter to Burghley, a text based entirely on verbal description and entirely without images.

If Ausonio's optics was crafted to fit the established format of the theoric, then what was Bourne trying to accomplish in his letter to Burghley? We have seen that Ausonio's optics was an appropriation of medieval optics, which selectively engaged with the optical tradition and adapted it in response to Ausonio's mathematical identity. I will suggest now that Bourne's optics is also a response to John Dee's and Leonard and Thomas Digges' appropriation of medieval optics, or, better,

[29]For Ausonio's instruments in the library of Emmanuele Filiberto, see Dupré, "The Dioptrics of Refractive Dials," 56–57. For the visual characteristics of Ausonio's theoric, see Dupré, "Visualization in Renaissance Optics," 26–33.

[30]See Ausonio, "Theorica speculi concavi sphaerici," in Galileo, *Le Opere di Galileo Galilei*, 3: 865.

to Bourne's perception of the *image* of optics developed in England at this time. It will become clear that Bourne hoped, although in vain, to attract Lord Burghley's patronage in this way.

William Bourne was a jurat, or town councilman, in Gravesend on the lower Thames.[31] In 1571–2 he served as the town's port-reeve, the equivalent of mayor. In Gravesend Bourne had everyday contact with sailors, and he also practised gunnery as a citizen volunteer at the defensive bulwark of Gravesend. Thus, although he was not university-educated, unlike some other mathematical practitioners in England in this period, the Gravesend context presumably established him on a career of writing on almanac-making, surveying, navigation and gunnery. How did Bourne come to write to Sir William Cecil (1520/1–1598), Principal Secretary and later Lord Treasurer to Elizabeth I, on practical optics and telescope design? The occasion for the letter was (Bourne wrote in his dedication) "that of late youre honour hathe had some conference and speache with mee, as concerning the effects and qualityes of glasses, I have thought yt my duty to furnish your desyer, according unto suche simple skill, as God hathe given me, in these causes".[32] Bourne was eager to attract Burghley's patronage and his letter on the optical properties and qualities of lenses and mirrors was not his first attempt to do so. In his letter Bourne reminded his potential patron:

> And allso aboute seaven yeares passed, uppon occasyon of a certayne written Booke of myne, which I delivered your honour, Wherin was set downe the nature and qualitye of water: As tuchinge ye sinckinge or swymminge of thinges. In sort youre Honoure had some speeche with mee, as touching measuring the moulde of a shipp. Whiche gave mee occasyon, to wryte a little Boke of Statick. Whiche Booke since that tyme, hath beene profitable, and helpped the capacityes, both of some sea men, and allso ship carpenters. Therfore, I have now written this simple, and breefe note of the effects, and qualityes of glasses, according unto the several formes, facyons, and makings of them ...[33]

This "certayne written Booke of myne, which I delivered your honour" was a manuscript, dedicated to Lord Burghley, that contained two works *Art of shooting in great Ordinance* and *Treasure for Travellers*—both of which were published in 1578—before an editorial decision was made to split them up.[34] As on the one hand

[31]On William Bourne, see G. L'E. Turner, "Bourne, William (c. 1535–1582)," *Oxford Dictionary of National Biography* (Oxford, 2004) [http://www.oxforddnb.com/view/article/3011, accessed 30 Aug 2005]. See also the introduction in E. G. R. Taylor, *A Regiment of the Sea and other Writings on Navigation by William Bourne* (Cambridge, 1963), xiii-xxxv; Taylor, *The Mathematical Practitioners of Tudor and Stuart England*, 33–39; E. G. R. Taylor, *Tudor Geography 1485–1583* (London, 1930), 155–156; E. G. R. Taylor, *The Haven-Finding Art: A History of Navigation from Odysseus to Captain Cook*, (London, 1958), 192–214; Samuel Bawlf, *The Secret Voyage of Sir Francis Drake 1577–1580* (New York, 2003), 68–73, 309–311.

[32]Bourne to Burghley, ca. 1580, in Van Helden, *Invention*, 31.

[33]*Ibid.*

[34]British Library (London), MS Sloane 3651. My appreciation of the dating of William Bourne's manuscripts is fully based on Stephen Johnston's unpublished and revised (with respect to Taylor's) bibliography of William Bourne. See http://www.mhs.ox.ac.uk/staff/saj/bourne/ [accessed 12 Sep. 2005]. Compare the bibliography of Bourne in Taylor, *Regiment*, 439–459.

Burghley was addressed as "Lorde Highe Treasurer of Engelande", a title he was awarded in the summer of 1572, and on the other, Bourne had announced these two works as ready for publication in his *Regiment for the Sea* (1574), the manuscript must have been written in 1572/3.[35] As "aboute seaven yeares passed", it follows that Bourne's letter on the "effects and qualityes of glasses" should be dated to 1579/80.[36] Between 1572/3 and 1579/80 Bourne had also written "a little Boke of Statick", a short hydrostatical text, for Burghley.[37] Moreover, the *Inventions or devices*—in which Bourne first mentioned a telescopic device—was published in 1578, but a version already existed in 1576 in a manuscript dedicated to Burghley.[38] Thus, Bourne's letter on the optical properties of mirrors and lenses came at the end of a decade in which Bourne had repeatedly sought Lord Burghley's patronage.

Burghley's interest in natural knowledge is to be considered in the light of his economic policy, which was marked by the development of the patent of monopoly as a means of advancing the commonweal.[39] This technique was also meant to encourage self-sufficiency by adopting and bringing foreign skill to England's economy. A nice case in point was the English glass industry: during the same period in which Bourne attempted to attract Burghley's patronage with a telescope design, the English glass industry experienced a revival due, above all, to the initiatives of the glassmaker Jean Carré, formerly of Antwerp, who brought with him glassmakers from Flanders, Normandy, Lorraine and Venice.[40] Burghley was much in need of natural knowledge related to glass-making when he had to review the applications for patents of monopoly.

A distinctive pattern of patronage of natural knowledge accompanied Burghley's economic policy with its focus on the patent of monopoly.[41] Burghley had a firmly utilitarian attitude to natural knowledge. His patronage was continually solicited by all kind of projectors who proposed all kind of inventions, and in some cases he was seduced into support of unlikely ventures, such as alchemical projects directed towards the transmutation of base metals. Even then, however, his interest stemmed from a utilitarian attitude towards natural knowledge, for his concern was with coinage. Thus, Bourne's letter suggesting the feasibility of a telescope design to

---

[35] *Ibid.*, 278.

[36] Johnston, *Revised Bibliography*. This dating differs significantly from Van Helden's (ca. 1585) and Turner's (ca. 1572). Compare Van Helden, *Invention*, 30; Turner, "Bourne".

[37] Bodleian Library (Oxford), MS Ashmole 1148, ff. 79–102.

[38] Lawrence J. Schoenberg Collection (private collection, University of Pennsylvania), ljs345.

[39] For Burghley's economic patronage, see Felicity Heal and Clive Holmes, "The Economic Patronage of William Cecil," in *Patronage, Culture and Power: The Early Cecils*, ed. Pauline Croft, 199–229 (New Haven & London, 2002). See also Michael A. R. Graves, *William Cecil, Lord Burghley* (London/New York, 1998), 149–168.

[40] Eleanor S. Godfrey, *The Development of English Glassmaking 1560–1640* (Oxford, 1975), 16–37. See also R. J. Charleston, *English Glass and the Glass Used in England, circa 400–1940* (London, 1985), 42–108; W. A. Thorpe, *English Glass* (London, 1949), 86–113.

[41] Stephen Pumfrey and Frances Dawbarn, "Science and Patronage in England, 1570–1625: A Preliminary Study," *History of Science*, 42 (2004): 157–160.

Lord Burghley was one among the many proposals that Burghley received in those years. It is unclear at present if and how Bourne's strategy of picking practical optics or Burghley's interest in questions of practical optics was related to the contemporary revival of the English glass industry.

Expertise in telescope design might have sounded useful to an English court which now and then was confronted with wandering charlatans who claimed to be able to make a telescope. For example, in April 1541 an Italian offered a telescopic device to Henry VIII. The French ambassador in London reported that 'there is an Italian here, aged about 70 years, who has shown this king that he would make a mirror and place it on top of Dover castle, in which mirror could be seen all ships that leave Dieppe. Although that seems incredible, he has persuaded this king to provide money to make it, and left yesterday to fulfil his promise'.[42] Belief in such fantastic claims could prove to be expensive, and expertise in optics that would have allowed a potential patron to make a distinction between the fantastic and the possible thus much desired.

Expertise in optics was hard to find in Elizabethan England, and continental developments in the discipline were late to arrive there.[43] In English mathematical practitioners' appropriation of the optical tradition, the figure of Roger Bacon was highly important.[44] In the sixteenth century Bacon gained the reputation as a powerful magician, renown based less on the use of demonic magic than on his grasp of mathematics. One of the most convincing advocates of this impression of Bacon was the English mathematician Robert Recorde, who wrote that

> … many thynges seme impossible to be done, whiche by arte may very well be wrought. And whan they be wrought, and the reason therof not understande, than say the vulgare people, that those thynges are done by negromancy. And hereof came it that fryer Bakon was accompted so greate a negromancier, whiche never used that arte (by any coniecture that I can fynde) but was in geometrie and other mathematicall sciences so experte, that he coulde dooe by theim suche thynges as were wonderfull in the syght of most people.[45]

The achievement of optical marvels played an important role in the fashioning of this image of Bacon as a magician. One of these optical marvels was the construction of an optical instrument that showed things that happened elsewhere or far away. Recorde continued:

---

[42]Marillac to Montmorency, 10 April 1541, cited in *The History of the King's Works*, ed. H. M. Colvin (London, 1982), 4:375. For the cultural references of this projected telescopic mirror, see Eileen Reeves, *Galileo's Glassworks: The Telescope and the Mirror* (Cambridge, Massachusetts 2008), 15–80. For sixteenth-century descriptions of telescopes (including Digges') as tools of espionage and similar instruments offered to Burghley, see also Jessica Wolfe, *Humanism, Machinery and Renaissance Literature* (Cambridge, 2004), 106–107.

[43]For perspective, see Christy Anderson, "The Secrets of Vision in Renaissance England," in *The Treatise on Perspective: Published and Unpublished*, ed. Lyle Massey (Washington and London, 2003), 323–347.

[44]A. G. Molland, "Roger Bacon as Magician," *Traditio*, 30 (1974): 445–460.

[45]Robert Recorde, *The Pathway to Knowledg, Containing the First Principles of Geometrie* (London, 1551), 8 (preface).

> Great talke there is of a glasse that he made in Oxforde, in whiche men myght see thynges that were doon in other places, and that was judged to be done by power of evyll spirites. But I knowe the reason of it to be good and naturall, and to be wrought by geometrie (sythe perspective is a parte of it) and to stande as well with reason as to see your face in a common glasse. But this conclusion and other dyvers of lyke sorte, are more mete for princes, for sundry causes, than for other men, and ought not to bee taught commonly.[46]

That such optical marvels were attributed to Bacon is no coincidence. Bacon had given more than enough reason for it in his own writings, most prominently in his 'Epistola de secretis operibus artis et naturae et de nullitate magiae', which contained references to such optical marvels.

> Glasses so cast, that things at hand may appear at distance, and things at distance, as hard at hand: yea so farre may the designe be driven, as the least letters may be read, and things reckoned at an incredible distance, yea starres shine in what place you please.[47]

In light of the importance of the figure of Bacon to the establishment of a *practical optics*, magic and wonder-making were as crucial to the status of practical mathematics or optics as were the practical and economical utility on which Lord Burghley placed so much stress.[48]

The English mathematical practitioners Leonard and Thomas Digges and John Dee read the optical tradition through the eyes of the figure of Roger Bacon as magician in order to make claims for the establishment of a practical optics in England. Dee was strongly influenced by the figure of Bacon, and even wrote a work about him, now no longer extant, in the 1550s.[49] It comes therefore as no surprise that in his *Mathematicall Praeface* (1570) to the first translation of Euclid in England, under the heading of "stratarithmetrie", or the military sciences, Dee hinted at the optical marvel of telescopic vision when he announced that the military man "may wonderfully helpe him selfe, by perspective Glasses. In which, (I trust) our posterity will prove more skillfull and expert, and to greater purposes, then in these days, can (almost) be credited to be possible".[50] In the same

[46] *Ibid.*, 8.

[47] Roger Bacon, "Epistola de secretis operibus artis et naturae ed de nullitate magiae," translation in *Frier Bacon his Discovery of the Miracles of Art, Nature, and Magick. Faithfully Translated out of Dr. Dees Own Copy, by T. M. and Never Before in English* (London, 1659), 20.

[48] For similarly heterogeneous attitudes toward automata between utility and wonder, see Alexander Marr, "Understanding Automata in the Late Renaissance," *Journal de la Renaissance*, 2 (2004): 205–222. For the rhetoric of utility of English mathematical practitioners, compare Katherine Neal, "The Rhetoric of Utility: Avoiding Occult Associations for Mathematics through Profitability and Pleasure," *History of Science*, 35 (1999): 151–178.

[49] Dee mentioned this work in the letter of dedication to Mercator, prefacing the *Propaedeumate aphoristica*. "The Mirror of Unity, or Apology for the English Friar Roger Bacon; in which it is taught that he did nothing by the aid of demons but was a great philosopher and accomplished naturally and by ways permitted to a Christian man the great works which the unlearned crowd usually ascribes to the acts of demons," *John Dee on Astronomy, Propaedeumate Aphoristica (1558 and 1568), Latin and English*, Wayne Shumaker, ed. and trans. (Berkeley, 1978),117.

[50] John Dee, The elements of geometrie of the most ancient philosopher Euclide of Megara ... With a very fruitfull praeface made by M. I. Dee (London, 1570), b.j.r.

Mathematicall Praeface Dee included "perspective" or the science of optics among the mathematical arts.

Likewise in *Stratiaticos* (1579) Thomas Digges presented his late father's optical and catoptrical studies and his invention of "perspective glasses" in connection with the figure of Bacon.

> As sithence Archimedes (Bakon of Oxforde only excepted) I have not read of any in Action ever able by meanes natural to performe ye like. Which partly grew by the aide he had by one old written booke of the same Bakons Experiments, that by straunge adventure, or rather Destinie, came to his hands, though chiefely by conioyning continual laborious Practise with his Mathematical Studies.[51]

In his *Pantometria* (1571), largely written by Leonard Digges, but prepared for publication by his son, the younger Digges again stressed the "demonstrations Mathematicall" at the basis of his father's "proportionall Glasses duely situate in convenient angles".[52] Digges and Dee foresaw the establishment of a practical optics, aimed at the making of marvellous instruments such as Bacon's alleged 'telescope,' devices that might also be useful for economical reasons or military purposes. This is not to say that all practitioners in England shared this image of optics—for example, Dee's interest in optics and mirrors also stemmed from his search for an astrological physics (not present in his *Mathematicall Praeface*) and this was an interest which Digges did not share.[53] However, it was Dee's and Digges' vision of a practical optics that provided the opportunity for a mathematician such as Bourne to claim expertise in a field of mathematical knowledge, *practical optics*, in an attempt to attract Burghley's patronage.

Bourne's development of a practical optics was almost certainly more empirical than Dee and Digges might have wished. It was based on his reading of Digges' and Dee's appropriation of the optical tradition rather than on any first-hand familiarity with the optical tradition. It was also more Bourne's response to their image of optics than the practical and more recognizably bookish optics that they had had in mind. It is important to note, therefore, that Bourne attempted to claim expertise in a kind of mathematical knowledge which would have allowed him to distance himself, both socially and intellectually, from craftsmen. In fact, Bourne aspired to

---

[51] Thomas Digges, *An arithmeticall militare treatise, named Stratioticos* (London, 1579), 189–190.

[52] Thomas Digges, *A Geometrical Practise, Named Pantometria, Divided into Three Bookes, Containing Rules Manifolde for Mensuration of All Lines, Superficies and Solides: With Sundry Straunge Conclusions Both by Instrument and without, and Also by Perspective Glasses, to Set Forth the True Description or Exact Plat of an Whole Region* (London, 1571), Aiiiv.

[53] For Dee's astrological physics, based on an optical model, see Nicholas H. Clulee, "Astrology, Magic, and Optics: Facets of John Dee's Early Natural Philosophy," *Renaissance Quarterly*, 30 (1977): 632–80; Idem, *John Dee's Natural Philosophy: Between Science and Religion* (London and New York, 1988), 39–74; Steven Vanden Broecke, "Dee, Mercator, and Louvain Instrument Making: An Undescribed Astrological Disc by Gerard Mercator (1551)," *Annals of Science*, 58 (2001): 226–9; Urszula Szulakowska, *The Alchemy of Light: Geometry and Optics in Late Renaissance Alchemical Illustration* (Leiden, Boston, and Köln, 2000), 12–70.

a social and intellectual status similar to that of John Dee and Thomas Digges. This desire is evident from Bourne's conclusion of his letter to Burghley:

> For that there ys dyvers in this Lande, that can say and dothe knowe much more, in these causes, then I: and especially Mr. Dee, and allso Mr. Thomas Digges, for that by theyre Learninge, they have reade and seene moo [*sic*] auctors in those causes: And allso, theyre ability ys suche, that they may the better mayntayne the charges: And also they have more leysure and better tyme to practyze those matters.[54]

Indeed, the difference in status between Bourne and Digges was also socially pronounced. Thomas Digges not only belonged to the class of gentry and landowners in Kent, but unlike Bourne, he also became a member of parliament with privileged access to members of the Privy Council such as Lord Burghley, whom he could count among his patrons.[55] Contemporaries likewise perceived the difference between Bourne and Digges in social as well as intellectual terms: Gabriel Harvey considered Bourne a "cunning and subtle Empirique," while Digges was placed on a par with "profound mathematicians" such as Thomas Harriot and Dee.[56] With his claim to expertise in practical optics, Bourne tried to win Burghley's patronage and to bridge the gap between his social and intellectual status and that of Digges. Insofar we can tell, in this and in all his other attempts, Bourne's application for Burghley's patronage went, however, unanswered.

7.4 Conclusion

In this essay, I have argued for recognition of the diversity of mathematical identities and images of optics in the sixteenth century. I have focused in particular on the different balances of theory and practice among Renaissance mathematicians' images of optics. Moreover, I have argued that this diversity is difficult to grasp in accounts of practical optics (and by extension, of practical mathematics) which oppose theory and practice as mutually exclusive categories. Rather than as an established category, practical optics should be understood as the result of a construction by Renaissance mathematical practitioners' appropriations of the perspectivist optical tradition. It is also as a consequence of these appropriative practices that significant divergences in the balance of theory and practice within practical optics emerged.

---

[54]Bourne to Burghley, ca. 1580, in Van Helden, *Invention*, 34.

[55]For Digges' biography and the shaping of his mathematical identity, see Stephen Johnston, "Digges, Thomas (c. 1546–1595)," in *Oxford Dictionary of National Biography* (Oxford, 2004) [http://www.oxforddnb.com/view/article/7639, accessed 30 Aug 2005]; Stephen Johnston, *Making Mathematical Practice: Gentlemen, Practitioners and Artisans in Elizabethan England* (Cambridge: PhD dissertation, 1994), 50–106. Digges was a client of Lord Burghley, to whom he had dedicated his "Alae seu scalae mathematicae," (1573), written in response to Burghley's queries on the nova of 1572. See *Ibid.*, 60–62.

[56]Gabriel Harvey, *Pierces Supererogation*, in *The Complete Works*, ed. Alexander Grosart, 3 vols., vol. 2 (London, 1884; rpt. New York, 1964), 289–290.

I have contrasted Ausonio's practical optics, which aimed at the design of an instrument by offering a theoric, with Bourne's project for the making of a telescope, which lacked any attempt at a theoric. Such differences in the balance of theory and practice are the consequence of variant readings of the optical tradition. While Ausonio's optics emerged in a direct but selective engagement with perspectivist optics, Bourne's optics developed without such direct contact with such sources. It emerged primarily as a response to what he saw as Dee's and Digges' image of optics, which in itself was based on a reading of the optical tradition derived primarily from Roger Bacon. His attempt to claim expertise in practical optics—as a type of knowledge that might differentiate him from the ordinary craftsman—was a bid for Burghley's patronage. That Bourne's patronage strategy failed is evidence of Burghley's perception that this differentiation was unsuccessful. The differences between Bourne's and Ausonio's practical optics which arose in response to the different demands imposed upon them in their respective contexts are, however, differences within the admittedly unstable category of practical optics rather than a clash between two endeavours of a different kind or a conflict within a community of 'mathematical practitioners'. It is Bourne's and Ausonio's shared knowledge that allows one to speak of a field of practical optics. This field was conceptually innovative vis-à-vis the optical tradition or the mixed mathematical field of optics to which it directly (in the case of Ausonio) or indirectly (in the case of Bourne) responded. It is perhaps ironic that the English case, of which—and rightly so—much is made in accounts that stress the role of practical mathematics and mathematical practitioners in the transformation of natural philosophy, is a dead end here. It was, nevertheless, the practical optics contributed to by Ausonio's appropriation of the mixed mathematical field of optics, as well as the optical tradition itself 'untouched' by practical optical concerns, that was available to natural philosophers, the likes of Kepler and Descartes.

Chapter 8
Hero of Alexandria and Renaissance Mechanics

W. R. Laird

Abstract The reception in the sixteenth century of the mechanical works of Hero of Alexandria offered an intriguing point of contact between humanists, mathematicians, engineers, and artisans. Although Hero's most important work, the *Mechanics*, was unknown in the West, Pappus of Alexandria had included Hero's theory of the five simple machines in his *Mathematical Collection*, whence it was adopted by Guidobaldo del Monte and Galileo as an organizing principle of theoretical mechanics. But in addition to the *Mechanics*, Hero wrote three other mechanical works: the *Pneumatica*, the *Automata*, and the *Belopoiica*, all of which were translated from the Greek and printed in the sixteenth century. Historians have suggested that these works generally encouraged experimental techniques and inspired an interest in mechanical technology.

This chapter traces the reception and influence of Hero's mechanical works through their manuscripts, their editions and Latin translations, and their wider dissemination among engineers and other practical men. Because the principles that govern pneumatic devices were not easily reconciled with the general principles of the other simple machines, such devices came to be classified as magic-working and thus contributed little to theoretical mechanics. And rather than inspiring the interest in practical machines, Hero's texts were studied as a result of the already existing interest in mechanical technology.

In their reception through the course of the sixteenth century, the mechanical works of Hero of Alexandria offered an intriguing point of contact between humanists, mathematicians, engineers, and artisans. Although Hero's most important work, the *Mechanics*, was apparently unknown in the West, surviving today only in an Arabic

An earlier version of this paper was read at the conference "The Mechanization of Natural Philosophy," Grenoble, France, 17–19 November 2005.

W.R. Laird (✉)
Carleton University, Ottawa, ON, Canada
e-mail: WRLaird@carleton.ca

L.B. Cormack et al. (eds.), *Mathematical Practitioners and the Transformation of Natural Knowledge in Early Modern Europe*, Studies in History and Philosophy of Science 45, DOI 10.1007/978-3-319-49430-2_8

translation made in the ninth century,[1] excerpts from it were included by Pappus of Alexandria in his *Mathematical Collection.* These included Hero's theoretical treatment of the five powers or simple machines, and from Pappus the five simple machines were adopted by Guidobaldo del Monte and Galileo, among others, as the organizing principle of sixteenth-century theoretical mechanics. But in addition to the *Mechanics*, Hero wrote three other mechanical works: the *Pneumatica*, which describes the principles and the construction of a number of machines operated by water and air pressure; the *Automata*, which describes the construction of two weight-driven puppet theatres; and the *Belopoiica*, which describes the construction of a crossbow and two catapults.[2] All three works are extant in Greek and all three were translated and printed in the course of the sixteenth century. Of these, the *Pneumatica* was by far the most important, to judge from the attention it has received both in the sixteenth century and more recently. Marie Boas, in an article published in 1949, sketched the history of the reception of the *Pneumatica* through the sixteenth century, and argued that the treatment of the interparticulate void in the introduction was an important source of non-atomistic matter theory in the seventeenth century.[3] Alex Keller, writing on the hydraulic engineer Giambattista Aleotti, argued that Hero's "chief contributions to the achievements of that generation [i.e., the early seventeenth century] lie in the experimental techniques which they derived from his instrumentation," and that his theories on matter and the void, opposed as they were to Aristotle's, were accepted through "the popularity of his pneumatic demonstrations."[4] My concern here, however, is with the influence of Hero's works not so much on theories of matter and the void in the seventeenth century but on the emerging science of mechanics in the sixteenth century. In particular, I would like to show that the reception of these works followed the general pattern of Greek mathematical and scientific writings in the Renaissance: the Greek manuscripts were first collected and copied by humanist scholars and bibliophiles, whence they came to the notice of mathematicians and were translated into Latin and the vernacular, and so received a much wider circulation. Secondly, because the principles that govern the hydraulic and pneumatic devices of the *Pneumatica* were not easily reconciled with the general principles of the other simple machines, rather

[1]On the *Mechanica*, see A. G. Drachmann, *The Mechanical Technology of Greek and Roman Antiquity* (Madison: University of Wisconsin Press, 1963); see also Romano Gatto, "La *Meccanica* di Erone nel Rinascimento," in *Scienze e rappresentazioni: Saggi in onore di Pierre Souffrin*, ed., Pierre Caye, Romano Nanni, and Pier Daniele Napolitani (Florence: Olschki 2015), which appeared too late to be incorporated into this article.

[2]On Hero's life and mechanical works, see A. G. Drachmann, "Hero of Alexandria," *Dictionary of Scientific Biography* 6, ed. Charles Coulton Gillespie (New York: Scribner, 1970), 310–314.

[3]Marie Boas, "Hero's *Pneumatica.* A Study of its Transmission and Influence," *Isis*, 40 (1949): 38–48, repr. in Marie Boas Hall, *The Mechanical Philosophy* (New York: Arno Press, 1981); on Hero's legacy in the seventeenth century for matter theory, including Galileo, see Boas, 46–48 (page numbers refer to the original article in *Isis*).

[4]A. G. Keller, "Pneumatics, Automata and the Vacuum in the Work of Giambattista Aleotti," *British Journal for the History of Science*, 3 (1967): 338.

than being incorporated into mechanics under a common set of principles, these devices came to be classified according to their main use—in thaumaturgy or magic-working—and thus contributed little to theoretical mechanics. And finally, although it is sometimes assumed that ancient technological texts such as Hero's inspired an interest in the design and construction of machines, I should like to suggest that generally the converse seems to have been true for Hero—that interest in his texts outside the circles of purely philological humanism was more the result than the cause of a general interest in mechanical technology. For by the time these texts were translated and thus widely available, the sorts of devices they described were either commonplace or obsolete. In fact, Bernardino Baldi, who translated both the *Automata* and the *Belopoiica* in the 1570s, apologized for the impracticality of these works and their lack of relevance to a modern age. In what follows, then, I should like to trace the reception and influence of Hero's mechanical works through their extant manuscripts and reported manuscripts, their editions and Latin translations by humanists and mathematicians, and their wider dissemination in vernacular translations by engineers and other practical men.

8.1 The Medieval Hero

In the Latin Middle Ages, Hero's works were almost completely unknown. The *Mechanics* is extant neither in the original Greek nor in Latin translation, and there is no unequivocal evidence of the survival of Greek manuscripts of it later than the ninth century, when Costa ben Luca translated it into Arabic. Nor does this Arabic translation seem to have had any direct or even indirect influence on mechanical thought or technology in the West. There is, however, evidence in the Middle Ages of Greek manuscripts of the *Pneumatica*, and perhaps of Latin translations now lost. Henricus Aristippus, in listing for a friend the sources of philosophy and science available in Sicily in 1156, says that "*habes Eronis philosophi mechanica pre manibus, qui tam subtiliter de inani disputat quanta eius virtus quantaque per ipsum delationis celeritas*," which description fits the *Pneumatica*. Valentin Rose suggested that Aristippus was referring to a Latin translation rather than the Greek text, which now seems unlikely, though Charles Homer Haskins suggested that a Latin version of the *Pneumatica* in a sixteenth-century manuscript in Paris, BN MS. Lat. 7226B, ff. 1–43, may be a copy of this lost medieval Latin translation.[5]Richard of Fournival, in the catalogue of his library written around 1250, mentions an "*excerpta de libro Heronis de specialibus ingeniis*"; Alexander Birkenmajer, emending *specialibus* to *spiritualibus*, concluded that this referred to a Latin translation of the *Pneumatica*, perhaps the same one mentioned by Aristippus. Edward Grant argued, however, that such titles more likely refer to Philo's work on pneumatics (which survived in a number of medieval

[5]Charles Homer Haskins, *Studies in the History of Medieval Science* (New York: F. Unger, 1925, rpt. 1960), 181n and 182.

manuscripts), though in a Cracow manuscript of Philo there is a fragment of what is perhaps again a medieval Latin translation of Hero's *Pneumatica*; the colophon of the fragment, however, cites its source as "*Ex libro Heronis de inani et vacuo*" rather than *De spiritualibus* or *specialibus*—so whether this is Fournival's excerpt or not remains in doubt.[6]

Again, Hero's *Pneumatica* was unmistakably referred to in Pseudo-Grosseteste's *Summa philosophiae* (*ca* 1270), though the source of the author's knowledge of Hero remains unknown. And Birkenmajer inferred the existence of a translation of Hero's *Pneumatica* by William of Moerbeke from the mention in 1274 of a book *De aquarum conductibus et ingeniis erigendis* that the arts faculty of the University of Paris claimed that Thomas Aquinas had promised them. Again, Grant has convincingly argued that this inference is implausible.[7] Moerbeke did, however, translate Hero's *Catoptrica* (which was then known as Ptolemy's *De speculis*) and provided a copy to his friend Witelo; the translation is included in the holograph of Moerbeke's translations of Archimedes now in Rome.[8] And finally, Giovanni da Fontana (1395?-1455?)—if he is indeed the author of the *Protheus*—mentions "a work of Hero de vacuo et inani," which must refer to the *Pneumatica* or at least to its introduction.[9]

8.2 Hero Among the Humanists

Despite these references and the possible identifications of fragments of medieval translations, the earliest extant Greek manuscripts and Latin translations of the *Pneumatica* date from the late fifteenth or early sixteenth century; many of them can be traced back to Venice, though a few are to be found in Florence and Rome, and there is evidence of others in Sicily and Naples.[10] Cardinal Bessarion, among his vast collection of Greek manuscripts, owned two copies of Hero, one an eleventh-century manuscript of the *Pneumatica* and *Automata* (Venice, Marciana MS. Z. Gr. 516), the other of the *Pneumatica* (now Venice, Marciana MS. Z. Gr. 263), which was copied into Laurenziana LXXXVI, 28.[11] Regiomontanus copied the *Pneumatica* from one or both of these into what is now Nuremburg MS. Cent. V, app. 6, which also contains copies of Aristotle's *Mechanica* and Apollonius'

[6]Edward Grant, "Henricus Aristippus, William of Moerbeke and Two Alleged Medieval Translations of Hero's *Pneumatica*," *Speculum* 96 (1971): 659–660.

[7]*Ibid.*, 662–669.

[8]Rome, Vat. Ott. Lat. 1850; Paul Lawrence Rose, *The Italian Renaissance of Mathematics* (Geneva: Droz, 1975), 80–81.

[9]Boas, "Hero's *Pneumatica*," 40.

[10]Rose notes that the library of the kingdom of Naples had in the fifteenth century a manuscript of Hero in Greek (Rose, *Italian Renaissance of Mathematics*, 55).

[11]Rose, *Italian Renaissance of Mathematics*, 35, 45.

Conica.[12] Regiomontanus also owned a Greek codex of the *Mechanica* and the *Pneumatica* (which is now lost and which he apparently collated with one or both of the Bessarion manuscripts). In his Tradelist of 1474—the list of books he intended to print—he included Hero's *Pneumatica* and something he called the *opus mechanicum mirae voluptatis* (*Mechanical Work of Marvelous Pleasure*), which is likely the *Automata.*[13]

Bessarion left his manuscript collection to the Marciana in 1468, where the sometime humanist and poet Pietro Bembo was later librarian. Bembo's private library contained a miscellany in Greek of works by Cleomedes, Hero, and Ptolemy, which is now in Rome.[14]

Manuscripts from Bessarion's library were later used by Diego Hurtado de Mendoza, the Imperial Ambassador to Venice, who borrowed the *Pneumatica* (MS. Z. Gr. 263) from the Marciana 1545–46.[15] A copy of the *Pneumatica* (Bruxelles, Bibliothèque Royale 3608) was made by a copyist working in the Venetian scriptorium that made many copies for Mendoza; the archetype was a manuscript now in Madrid (Escorial Φ . I. 10), which according to its colophon was itself written in Venice in 1542 and which was also the archetype of Paris, BN Ancien fonds grec 2428; Paris 2428 is listed in the 1543 catalogue of Mendoza's library.[16] Mendoza, who translated the *Mechanical Problems* into Spanish in 1545 while attending the Council of Trent, studied mechanics with Niccolò Tartaglia in Venice and appeared as an interlocutor in his *Quesiti ed inventioni diverse* of 1546.[17] Tartaglia, according to his *supplica* for a copyright of 1542, intended to print, among other works, his own translation of Hero, which apparently he never made or published.[18] Perhaps Tartaglia planned to use a manuscript provided by Mendoza.

8.2.1 Giorgio Valla

In his *Miscellaneorum Centuria Prima* (1489), the humanist and poet Angelo Poliziano drew on several Greek mathematicians, including Hero, and chapter 97 describes various automata found in Hero's *Pneumatica*; in his *Panepistemon* (1490–91) he cites Hero's *Automata* and *Pneumatica*. His sources were perhaps

[12] *Ibid.*, 99.

[13] *Ibid.*, 105–107.

[14] Rome, MS. Vat. Gr. 1411 (Rose, *Italian Renaissance of Mathematics*, 11).

[15] Rose, *Italian Renaissance of Mathematics*, 46.

[16] Ole L. Smith, "On Some Manuscripts of Heron, *Pneumatica*," *Scriptorium*, 27, no. 1 (1973): 96–101.

[17] Paul Lawrence Rose and Stillman Drake, "The Pseudo-Aristotelian *Questions of Mechanics* in Renaissance Culture," *Studies in the Renaissance*, 18 (1971):86–88.

[18] Rose, *Italian Renaissance of Mathematics*, 152.

the manuscripts that can be associated with Florence at the time, including the Greek Hero that was in the library of the Dominican convent of San Marco in 1500 and that may have belonged to Niccolò Niccoli, and another, unlisted in the 1495 inventory of the Libraria Medicea Privata (predecessor to the Laurenziana), that was noted as having been "returned to Cardinal Giovanni de' Medici in Rome in 1510." But a major nexus of Hero manuscripts seems to have been the Venetian humanist and encyclopedist Giorgio Valla. Poliziano noted that on a visit to Valla's house at Venice in June 1491, he found "some mathematical books of Archimedes and Hero that we lack," which Rose suggested was probably the *De mensuris*, since that same year Poliziano requested Valla's copy of the *De mensuris* for copying. The *De mensuris* was bound with the works of Archimedes in the famous Codex A, owned by Valla at the time; and the copy of Codex A made for Lorenzo de' Medici (Laurenziana MS. XXXVIII, 4) in fact contains the *De mensuris*. Janus Lascaris, who was sent by Lorenzo de' Medici to look for Greek manuscripts, saw a copy of Hero's *Pneumatica* in Valla's library in 1490, and also owned a copy himself.[19]

But Valla's greatest contribution to the knowledge and dissemination of Hero's *Pneumatica* was his inclusion of a large excerpt translated into Latin in his encyclopedic *De expetendis et fugiendis rebus opus* of 1501.[20] Valla introduced the excerpt by noting that pneumatics is a part of mechanics and thus belongs to the mathematical sciences, and that it depends on the principle that nature does not admit a void, a principle that gives rise to various marvelous motions. But since the examples of machines that Hero described belong more to the senses than to the soul, Valla deferred the question of the void to a discussion of physics; as a result, he omitted the entire introductory part, in which Hero discussed his theory of matter, the interparticulate void, and the creation of artificial vacuums by art and force, which are the principles of the devices that followed (f. zvii verso). Valla followed this brief introduction with Latin translations or very close paraphrases of 24 of Hero's 78 descriptions of pneumatic machines, including siphons, liquid dispensers, the wind-powered organ, the aeolipile, and the sun-powered fountain. Valla omitted all of the animated figures, the lamps, and all but one of the temple devices; perhaps he considered them frivolous, redundant, or irrelevant. The figures that accompany Valla's text are very close to those of BM MS. Burney 108, which Boas in her note to Woodcroft states is a faithful copy of Marcianus 516, the oldest extant Greek manuscript.[21]

Valla's excerpt was the only printed version of Hero's *Pneumatics* for 75 years, and seems to have been the main source for Hero until the editions and translations of the 1570s, 80s, and 90s. Leonardo da Vinci, for one, probably got his knowledge

---

[19]Rose, *Italian Renaissance of Mathematics*, 32–35, 47, 63n90.

[20]Giorgio Valla, *De expetendis de fugiendis rebus opus* (Venice: Aldus, 1501), Book XV, fols zvii verso – aaiii verso.

[21]Marie Boas Hall, "A Note on the Text and its Accompanying Illustrations," in Hero of Alexandria, *The Pneumatics of Hero of Alexandria*, tr. for and ed. by Bennet Woodcroft (London, 1851); rpt. in facsimile with Introduction and Note by Marie Boas Hall (London: MacDonald, 1971), 119.

of Hero's devices from Valla and, if so, it is not surprising that he was not influenced by Hero's theory of matter (as W. Schmidt observed), since Valla omitted that part of Hero's text.[22] And in 1567, the engineer Giuseppe Ceredi noted that he had acquired a copy of Pappus and of Hero that had come from Valla's library in Venice.[23]

8.2.2 Latin Translations

There were several Latin translations of the *Pneumatica* made in the mid-sixteenth century, and although they circulated in manuscript, none was printed. According to correspondence now in the Vatican, Cardinal Marcello Cervini, who later became Pope Marcellus II, translated at least part of Hero's *Pneumatica* into Latin in 1533 from a Greek manuscript he borrowed from his friend, the Roman humanist and antiquary Angelo Colocci, and which later came into the possession of Fulvio Orsini, librarian of the Biblioteca Farnesiana under Ranuccio Farnese (1530–1565), and which is now in the Vatican Library (Vat. Gr. 1364). Cervini's translation, however, is apparently not extant.[24]

Giovanni Battista Gabio (Johannes Gabius), professor of Greek at the Roman Sapienza, translated the *Pneumatica* into Latin under the title of *Spiritualia*. This translation is extant in two manuscripts that I know of—MS. Barberiniano latino 310 (X, 128) and Vat. Lat. 4575.[25] More widely known, it seems, was the translation made by Francesco Burana of Verona, *De spiraminibus* [or *spirabilium*] *liber primus*, which is extant in four manuscripts that I have found.[26] The copy in Rome

[22]W. Schmidt, "Leonardo da Vinci und Heron von Alexandria," *Biblioteca Mathematica* 3, no. 3 (1902), 180–187, cited in Boas, 40–41 and 41n.

[23]Rose, *Italian Renaissance of Mathematics*, 47.

[24]Cervini also possessed a copy of Hero's *De ponderibus et mensuris* (since Colocci asked to borrow it) and a copy of Hero's *De geometria*, which is listed in an inventory of his manuscripts (in Rome, Vat. Lat. 8185); Cardinal Sirleto had a copy of Hero's *De geometria* made for Cervini in 1546–47 from a manuscript in Perugia—all this is from correspondence in Rome, Vat. Lat. 6177, fols 30, 32, 119, 320, and Vat. Lat. 6178, fols 150, 130 (Rose, *Italian Renaissance of Mathematics*, 190–191).

[25]The first is listed in Paul Oskar Kristeller, *Iter Italicum*, (London; Leiden, 1965–67), II: 444 — see Elio Nenci, ed., Girolamo Cardano, *De subtilitate* Vol. I: Books I-VII (Milan, 2004), 61n-62n; for the second, see Rose, *Italian Renaissance of Mathematics*, 50.

[26]Paris, Bibliothèque Nationale, Fonds Latin MS. 10261, and Rome, Biblioteca Lancisiana MS. 249, are listed by Kristeller in *Iter Italicum* II, 118b; Milan, Biblioteca Ambrosiana MSS. J 38 inf. and G 78 inf. by Nenci, *Cardano*, 61n. Milan, Biblioteca Ambrosiana MS. N 237 sup. also contains what appear to be an Italian version or derivitive of the *Pneumatica – Degli effetti de' venti ed invenzioni curiose di meccanica, getti di acqua e fontane, ecc.* (fols 1–50r)—and an Italian translation of the *Automata – Trattato delle cose che si muovono* (beginning on f. 56)—as listed in Astrik L. Gabriel, *A Summary Catalogue of Microfilms of One Thousand Scientific Manuscripts in the Ambrosiana Library, Milan* (Notre Dame, Ind.: The Mediaeval Institute, 1968), no. 686, and Paolo Revelli, *I codici Ambrosiani di contenuto geografica* (Milan: Luigi Alfieri, 1929), no. 267, 102.

is accompanied by fine drawings, but it has no preface, commentary, or notes that might provide evidence of the circumstances surrounding or the reasons for the translation.

8.2.3 Girolamo Cardano

One of these translations was probably used by the physician, mathematician, and polymath Girolamo Cardano—unless he used a Greek text—to supplement Valla, which was perhaps his main source. The first book of Cardano's *De subtilitate* (first published in 1550) was devoted to the general principles of "matter, form, vacuum, the repulsion of bodies, natural motion, and place;" his discussion of the various devices that some claim prove the existence of the vacuum are derived in part from Valla. But Cardano must have had a source beyond Valla, since he described a machine that Valla had omitted (*Pneumatica* I.37). Elio Nenci, Cardano's modern editor, suggested that he also relied on the preliminary discussion of the vacuum omitted by Valla, but I see no evidence of that—Cardano never mentions the interparticulate vacuum, for instance.[27]

From the first Cardano argued that a vacuum does not and cannot exist. Rarity and density, he says, concern only the quantity of material; and since all material has existed from the beginning and fills the sphere of the world, there could not have been a vacuum, and to create one some material would have to be destroyed. He then asserted that there were many reasons that show that a vacuum cannot exist, and although Hero in the *Spiritualibus* had tried to show its existence, "it is not for the wise to refute every absurdity." The bellows, for instance, show that a vacuum cannot be created: for if bellows are forced open without allowing air to enter, they will break before a vacuum is made. Similarly, this is why water will rise from a pitcher when sucked through a tube. He then described a lamp that is fed by a large reservoir, which lets oil flow only when the level at the wick allows air to enter.[28]

According to Cardano, the motion apparently caused by the vacuum is natural, since it is caused by the form of the element, which does not admit of greater rarity or allow its parts to be separated. "Prime material does not admit separation," he asserts. Such motion is natural because it arises from this universal property of matter, even when it makes heavy things rise and light fall. Similarly, the motion that results from the impenetrability of material is also natural; as examples he describes

[27]Nenci also argues that the title Cardano gives ("*Spiritualibus*") suggests the version by Giovanni Battista Gabio rather than that by Burana (Nenci, *Cardano*, 61n).

[28]Cardano, *De subtilitate*, ed. Nenci, 61–63; the bellows were often cited in the sixteenth century as disproving the void. See Charles B. Schmitt, "Experimental Evidence for and against a Void: The Sixteenth-Century Arguments," *Isis* 58 (1967): 352–366, repr. in Charles B. Schmitt, *Studies in Renaissance Philosophy and Science* (London: Variorum Reprints, 1981).

Ctesibius's pump from Vitruvius X. 7, a ship's pump made by Bartholomaeus Brambilla, which Cardano saw in Milan, and finally a water clock.[29]

From these premises, Cardano then explains the siphon: water rises in the siphon and flows out only if the end is lower than the level of the water in the container, not because of the pull (*tractatio*) of the water (since it does not matter whether water, wine, oil, or milk fills the vase), but because the water in the reservoir above the level of the mouth of the siphon is lighter than the water at the mouth and because it desires to descend below the mouth, it therefore presses on the water and pushes it into the tube. The water in the vase below the level of the mouth, on the other hand, does not desire to be at the mouth, since it is already below it, and so does not want to rise. Therefore the water is not caused to rise in the tube because it is drawn by the water flowing out, but rather because it is pushed by the water within the container helped by the air entering it.[30] This is actually not a bad explanation, except that it does not explain why the water rises when one sucks on the tube. And Cardano's denial of cohesion as the cause of the action of siphons is contradicted by his earlier appeal to the mysterious property of prime matter to resist separation.

8.2.4 Francesco Maurolico

One of the Greek mathematical texts the Sicilian mathematician Francesco Maurolico intended to edit, according to the 1540 letter of dedication to his *Cosmographia* (1543), was Hero's *Pneumatica*.[31] Several folios of fragments dated 25 October 1534 of this projected edition are listed under the heading *Ex Heronis et aliorum spiritalibus* in a manuscript now in Rome.[32] And in the sketch of an encyclopedia, perhaps intended for teaching at the Jesuit College in Messina, Maurolico included in Book 11, on the mechanical arts, a section called *De machinis hydraulicis quaedam notanda*, which would have perhaps relied at least in part on Hero.[33] Unfortunately, neither the edition nor the encyclopedia was ever completed.

[29] Cardano, *De subtilitate*, ed. Nenci, 66–73.

[30] *Ibid.*, 75.

[31] Rose, *Italian Renaissance of Mathematics*, 163–164.

[32] Biblioteca Nazionale Vittorio Emanuele, MS. S. Pantaleo 115, fols 45-46v, according to Rose, *Italian Renaissance of Mathematics*, 161; San Pant. 115/32, fols 43r-44v, 46r-47v, according to Pier Daniele Napolitani, "Mechanicae artes," in *Francisci Maurolyci Opera Mathematica*, eds. Veronica Gavagna and Pier Daniele Napolitani (Pisa: Dipartimento di Mathematica dell'Università di Pisa, 2002), 91. On these fragments and their sources, see W.R. Laird, "The Sources of Francesco Maurolico's *Ex Heronis et aliorum spiritalibus*," *Bollettino di Storia delle Scienze matematiche* 30, no. 1 (2010): 9–21.

[33] Rosario Moscheo, *Francesco Maurolico tra Rinascimento e scienza galileiana* (Messina: Societa messinese di storia patria, 1988), 544.

8.2.5 Giuseppe Moletti

Though a student of Maurolico's in Messina, Giuseppe Moletti (1531–1588) does not seem to have learned anything from him about Hero or about mechanics in general. Moletti, a physician and mathematician, spent some time in Venice before he was taken on as mathematics tutor with the Gonzago court in Mantua in 1570. It was around this time that he wrote his *Discourse on the Nature of Mathematics*, a kind of professional manifesto in praise of the usefulness of mathematics to princes.[34] In the section on mechanics, he distinguished two main principles on which all mechanical effects depend: the circle (that is, the principle of circular movement that he found in the Pseudo-Aristotelian *Mechanical Problems*), and air or the vacuum.[35] The air and the vacuum are not two different principles, he says, but one, since the one works by the lack or absence of air, the other by its superabundance or multiplication. As examples of the power of the lack of air – which he attributes to nature's abhorrence of the vacuum – he notes the difficulty of extracting a post from a hole in the ground, and the difficulty of separating sheets of marble. As examples of the power of the superabundance or forceful multiplication of air, he cites organs and flutes, mines and artillery.[36] He then describes the clepsydra used to water gardens, which holds or releases water by stopping or allowing air to enter from the top. He also describes pumps used to lift water out of ships and from wells. And he describes how water can be made to rise in a flask that has been first heated and then inverted in a container of water: since the air in the flask is rarified when it is heated, when it cools it condenses, drawing the water up into the flask. Finally, he mentions the destructive force of the multiplication of air, demonstrated spectacularly in the explosion at the Venetian Arsenal in 1569, which damaged houses and palaces, scattering debris more than half a mile and breaking windows, the blast of which could be heard 25 miles away. In none of this does he show any familiarity with the works of Hero or even with Valla (whom he does not list among his sources for mechanics); his examples (except for the explosion in the Arsenal) could all have been found in Vitruvius. At the very end, however, he mentions that there are some books by Hero that he has heard are to be translated by the Venetian Senator Francesco Barozzi.[37]

But by 1581, when he first lectured on the *Mechanical Problems* at the University of Padua, Moletti was better informed about Hero. In those lectures he divided mechanics along the same lines as suggested in the *Discorso*, but with considerably more detail: of machines that work by air he distinguished those that work without

[34]Giuseppe Moletti, *Discorso che cosa sia matematica*, Milan, Biblioteca Ambrosiana MS. S 103 sup., fols 122r–175v.

[35]Moletti, *Discorso*, fol. 156r–v.; Moletti also mentions the two principles in a letter to Pinelli dated at Venice 18 June 1568, calling the principle of the circle mathematical, the principle of the vacuum natural (Milan, Biblioteca Ambrosiana MS. D 191 inf., fols. 52r–53v, on 52v).

[36]Moletti, *Discorso*, fol. 162r–v.

[37]*Ibid.*, fols. 165v–167r.

the multiplication of air but that make sounds, such as organs, horns, and flutes, "and the many machines that are mentioned by Hero in the book *De spiritualibus*." Those that do not make sounds include the hydraulic machines of Ctesibius, the Archimedean screw, ships' pumps, and bellows pumps used to inflate balls. Again, he mentions Vitruvius, who called all such machines pneumatical, that is *spiritales*.[38]

In the course of these lecture notes Moletti mentioned Hero's *De spiritualibus* again, in his list of mechanical authors. But there he made an interesting observation: he relates that he himself tried to make one of the fountains Hero described—though he does not say which or what his source for it was. When the fountain failed to work properly, he sought the advice of an artisan experienced with such work. The artisan told him that the pipes were not made properly, and when the artisan had corrected this defect with his special knowledge, the fountain then worked as it was supposed to. From this Moletti concluded that Hero "was content with demonstration and invention, leaving to the artisans those things that concern the material construction."[39] What is interesting here is that Hero's text seems to have prompted Moletti to try to build one of the machines—but notice: there were already at hand artisans experienced in such work, who presumably were already building these machines long before amateurs like Moletti, inspired by ancient texts, tried their clumsy hand at it.

Later in the lectures, in the section on mechanical authors, Moletti listed both the *Pneumatica* (*De spiritalibus*) and the *Automata* (*De his que per se moventur*), noting that only the first had been translated into Latin, by Federico Commandino, and printed. He also mentioned Hero's *De machinis bellicis*, which had been printed and translated into Latin by Francesco Barozzi, though he follows Barozzi in attributing this work to a Hero of a much later date – presumably this is the translation by Barozzi that Moletti alluded to in the *Discorso*. More importantly, he knew of Commandino's unprinted Latin translation of Pappus's *Mathematical Collection*, though he seems not to have been influenced by it.[40]

8.2.6 Federico Commandino

The 1570s and 80s finally saw the translation of Hero's extant mechanical works, a labor begun by the mathematician and editor and translator of Greek mathematical texts Federico Commandino (d. 1575) and continued by his student, the mathematical biographer Bernardino Baldi (d. 1617). Commandino had connections with

[38]Giuseppe Moletti, *In librum Mechanicorum Aristotelis expositio*, Milan, Biblioteca Ambrosiana MS. S 100 sup., fols154r–210v, on fol. 172r.

[39]"ipse fuit contentus demonstratione et inventione, relinquens ea. quae ad materialem constructionem attinebant artificibus" (Moletti, *Expositio*, fol. 184r).

[40]Moletti, *Expositio*, fol. 184r–v.

some of the leading humanist bibliophiles of the mid-century, including Cardinal Cervini and his mathematics teacher at Urbino, Pietro de' Grassi, who was a friend of Cardinal Niccolò Ridolfi, who possessed some 600 Greek manuscripts, including Hero's *Pneumatica*.[41] At Urbino, Commandino used a manuscript of Hero's *Pneumatica* and *Automata* (Urb. Gr. 75), and by 1572 he had already begun to translate into Latin Hero's *Pneumatica* and Pappus's *Mathematical Collection*, for in the letter of dedication to his edition of Euclid printed that year he stated that they were in progress, and they are included among the works for which he was granted a copyright privilege by Gregory XIII (the privilege is printed by him in the Euclid edition). Unfortunately, Commandino did not live to see his Hero and Pappus into print. His Latin translation of the *Pneumatica* was printed only after his death by his son-in-law, Valerio Spaccioli, in 1575.[42] As for the Pappus, the Greek manuscripts that Commandino worked from were missing Books I and II and contained a defective Book VIII (the mechanical book); there are two extant copies of Commandino's version of Book VIII: Marciana MSS. Z. Lat. 330 (1987) and 331 (1761), both of which had been owned by the Venetian mathematician Jacomo Contarini. After Commandino's death, Barozzi undertook to revise the translation with the aid of a Greek manuscript of Books II-VIII owned by Pinelli, the Venetian bibliophile. In all, Pinelli owned Greek manuscripts of seven works or fragments of works by Hero. Guidobaldo del Monte then printed the revised translation at Pesaro in 1588.[43] But as I mentioned before, Pappus had already exerted his influence on mechanics, since Guidobaldo had structured his *Liber mechanicorum* of 1577 according to the five powers or simple machines of Hero as found in Book VIII of Pappus's *Mathematical Collection*, and he took over Pappus's mistaken account of the inclined plane in preference to the correct one found in the medieval science of weights.[44]

8.2.7 Bernardino Baldi

Bernardino Baldi (1553–1617) set out to complete the work of Commandino, the man whom he says he loved as a father. More literary humanist and historian of mathematicians than mathematician himself, Baldi nevertheless contributed the diagrams to Commandino's translations of Pappus and Hero's *Pneumatica*, edited the Greek *Belopoiica*, translated into Italian both the *Automata* and the *Belopoiica*, and composed a Life of Hero as one of his series of Lives of the Mathematicians. The *Automata* he translated from Greek into Italian in 1576, at first using a poor

[41] Rose, *Italian Renaissance of Mathematics*, 186.

[42] *Ibid*., 205, 208.

[43] *Ibid*., 210–13; Marcella Grendler, "A Greek Collection in Padua: The Library of Gian Vincenzo Pinelli (1535–1601)," *Renaissance Quarterly*, 33 (1980): 402–405.

[44] Rose, *Italian Renaissance of Mathematics*, 224.

Greek manuscript owned by Commandino and then a better one obtained from Pinelli; revised, it was printed in Venice in 1589.[45] In the introduction to the printed edition Baldi described mechanics as a science subalternated to mathematics, part of which is concerned with making machines that move themselves (automata), some of these being moved by air or the vacuum, others by weights. Then he gives a brief history of automata, from Homer, through Vulcan and Daedalus, mentioning Architas's dove, Aristotle's *De motu animalium* (where the heart is compared to an automaton), Archimedes's spheres, Ctesibius, and Hero. More recently, he mentions mechanical clocks, the mechanical eagle and mechanical fly attributed to Regiomontanus, and most notably, a walking mechanical tartar made by Bartalomeo Campi (c. 1525–1573) for "our prince" (presumably the Duke of Urbino). Automata, it would seem, were not entirely unfamiliar in Urbino. The introduction concludes with an explanation of the stories acted out by the two puppet theatres.[46]

Baldi also edited the Greek and translated Hero's *Belopoiica* from a Greek manuscript obtained from his teacher of Greek at Padua, Manuel Maximos Margunios; his edition of the Greek is annotated with variant readings from a Vatican codex ("Cod. Vat." in the notes to the Greek). According to the autograph, the translation was completed in 1612; it was printed at Augsburg in 1616.[47] In his Life of Hero printed in the same volume, after surveying the sources for Hero's life and works, Baldi concluded with a discussion of vacuum. He noted that Hero's account contradicts Aristotle and agrees rather with Democritus and Epicurus, quoting passages from Lucretius in support. He was careful to note that Hero held that the only vacuum that occurs by nature is interparticulate.[48] Baldi, then, seems to have been the first author on mechanics that took seriously Hero's theory of matter and void and that did not dismiss it out of hand with arguments taken from Aristotle.

In addition, Baldi translated Book VIII of Pappus's *Mathematical Collection*, the book on mechanics that contains the excerpt from Hero's *Mechanics* on the five simple machines. This translation was made in 1578, before Commandino's

---

[45]*Di Herone Alessandrino degli Automati, overo Machine se moventi Libri due tradotti dal greco* (Venice, 1589; rpt. 1601, 1661, and reportedly 1647; rpt. Milan: Restampe Anastatiche, 1962); the autograph of the translation is in Florence, Laurenziana, MS. Ashburnham 1525 (Rose, *Italian Renaissance of Mathematics*, 246); Baldi mentions the Greek texts and his sources for them in his Preface, fols 3r–4v.

[46]Baldi, *Di Herone Alessandrino degli Automati*, fols 4r-16r; for a paraphrase of this Introduction, see Alessandra Fiocca, "Giambattista Aleotti e la 'Scienza et arte delle acque,'" in *Giambattista Aleotti e gli ingegneri del Rinascimento*, ed. Alessandra Fiocca (Florence: Olschki, 1998), 53; see also Rose, *Italian Renaissance of Mathematics*, 247, and Enrico Gamba and Vico Montebelli, *Le scienze a Urbino nel Tardo Rinascimento* (Urbino: Quattro Venti, 1988), 20n–21n.

[47]Hero of Alexandria, *Heronis Ctesibii Belopoeeca, hoc est Telifactiva*, tr. Bernardino Baldi (Augsburg, 1616); the autograph is in Paris, BN, Lat. 10280, fols 2-30v (Rose, *Italian Renaissance of Mathematics*, 247).

[48]Bernardino Baldi, *Heronis vita eodem auctore, in Heronis Ctesibii Belopoeeca, hoc est Telifactiva* (Augsburg, 1616), 74–76.

had been revised or printed, though after Guidobaldo had incorporated the simple machines and Pappus's account of the inclined plane into his *Liber mechanicorum* of 1577.[49]

8.2.8 Italian Translations of the Pneumatica

Several translations or partial translations of the *Pneumatica* appeared in the 1580s and 90s, made by engineers and other non-mathematicians. The engineer Vanocchio Biringucci made an Italian translation of the *Spiritalia* in 1582 and dedicated it to the painter and architect Bernardo Buontalenti, the presentation manuscript of which is now in Siena.[50] And in the same year the same Buontalenti was the recipient of a translation of the Preface only of the *Pneumatica*, under the title of *Volgarizzamento della natura del vuoto di Herone*, by the Florentine merchant and translator of Tacitus, Bernardino Davanzati (1529–1606), the manuscript of which is now in Florence.[51] Both these translations were probably made from Commandino's Latin translation of 1575, and neither, as far as I know, was ever printed.

The first Italian translation of the *Pneumatica* to be printed (in 1589) was made by the hydraulic engineer Giambattista Aleotti, which, according to the letter of dedication to Duke Alfonso II d' Este, Aleotti undertook in 1586 while convalescing and unable to do any practical work.[52] Although Aleotti added some theorems of his own, there was little in Hero that could be of use to him in his professional capacity. The only mention that Hero made of large-scale works was that a large siphon could not be started by sucking with the mouth (Theorem 5). In his *Hidrologia*, Aleotti mentioned Hero in his own explanation of how to start a large siphon to raise water over a dike, not by sucking, but by stopping the lower end and filling the siphon from a hole at the top of its arc.[53] Although it has been suggested that in ancient authors such as Hero, Aleotti and others "hoped to find the paradigm of a new science of making *(del fare)*,"[54] it would seem that they had already found this new paradigm without him.

[49]Baldi's autograph of Book VIII is in B.N.P. MS. Lat. 10280, fols 183ff, signed and dated 7 April 1578 on fol. 202 (Rose, *Italian Renaissance of Mathematics*, 213, 247).

[50]Siena, Biblioteca Comunale degli Intronati, MS. L.VI.44 (Daniela Lamberini, "Cultura Ingegneristica nel Granducato di Toscana ai tempi dell' Aleotti," in *Giambattista Aleotti*, ed. Fiocca, 306–307 and 307n).

[51]Florence, Biblioteca Nazionale Centrale, B. R. 223 (ex Palatino 1166, E.B.5.1.9), fols 1–8 (Lamberini, "Cultura Ingegneristica," 307 & n; and Boas, "Hero's *Pneumatica*," 41).

[52]*Gli artificiosi et curiosi moti spirituali di Herone*, Italian tr. Giambattista Aleotti (Ferrara: Baldini, 1589) (Rose, *Italian Renaissance of Mathematics*, 247; and Fiocca, "Giambattista Aleotti," 51–52).

[53]Fiocca, "Giambattista Aleotti," 54–56; see also 59–61 for other citations of Hero.

[54]Vittorio Marchis and Luisa Dolza, "L'Acqua, i numeri, le macchine: al sorgere dell' ingegneria idraulica moderna," in *Giambattista Aleotti*, ed. Fiocca, 210–211.

A second printed Italian translation of the *Pneumatica* was made by Alessandro Giorgi, an acquaintance, or perhaps a pupil, of Commandino's at Urbino, and printed in Urbino in 1592 (and reprinted in Venice in 1595) accompanied by a critical apparatus, a commentary on the devices, and references to contemporaries.[55] Giorgi also gives a brief life of Hero based on Baldi's (fol. 1r–v); in his introduction he describes the subject of the book as that part of mechanics called thaumaturgy (magic-working); he then gives a general account of the principles of motion that it uses and a brief discussion of the vacuum, the existence of which he dismisses on Aristotle's grounds (fols. 2r-6v). Giorgi's was the translation that Galileo possessed.[56]

Finally, this relegation of Hero's mechanical works to thaumaturgy was confirmed by the inclusion of a section on pneumatics in the 1589 edition of Giambattista della Porta's *Magia naturalis*, which first appeared in 1558; the addition was based largely on Hero, presumably from Commandino's translation. Della Porta also printed his own Italian translation of the *Pneumatica* in 1601.[57]

8.3 Conclusion

By the end of the sixteenth century, Hero's then-extant mechanical works—the *Pneumatica*, the *Automata*, and the *Belopoiica*—were all available in print in Latin or Italian translations. What can we now say of their place in sixteenth-century mechanics?

Bernardino Baldi perhaps expressed most clearly the general attitude towards these works of Hero in the latter part of the century. In the letter dedicating his edition and translation of the *Belopoiica* to Laelio Ruino, Bishop of Bagnorea (Balneoregiensis), Baldi explained the reasons for publishing this text (recall that the *Belopoiica* described the construction of a crossbow and two catapults) in these words:

> And so, although some of my friends not only encouraged but even urged and impelled me to publish this work, nevertheless I set my mind with difficulty to listen to them, for I knew that the very topic is vain today and that, after the invention of bronze artillery (or should I say monsters or lightning bolts), it would bring little of use or importance to military affairs. While I considered this, it occurred to me that without doubt the work of those who, with all their heart, try to shed light on ancient and noble learning and draw forth from the shadows the best inventions of ancient men, whatever they may be, is worthy to be praised

[55] *Spirituali di Herone Alessandrino ridotti in lingua volgare*, tr. Alessandro Giorgi (Urbino, 1592; rpt. Venice, 1595) (Boas, "Hero's Pneumatica," 42; Gamba and Montebelli, *Le scienze a Urbino*, 23).

[56] Boas, "Hero's Pneumatica," 48; Rose, *Italian Renaissance of Mathematics*, 281; on Galileo's library see Antonio Favaro, "La libreria di Galileo Galilei," in *Bullettino di Bibliografia e di Storia delle Scienze Mathematiche e Fisiche*, 19 (1886), 219–290.

[57] Boas, "Hero's *Pneumatica*," 43.

> and advanced. For although the writings of the illustrious ancients in this field are in fact not useful or opportune, nevertheless no-one sane would deny that these old things are, or can be, the stimulus of great thoughts for modern ingenious men. All these very arts, not only the noble but also the sedentary, thrive today and have recovered almost their original beauty. Therefore it was not right that the knowledge of the most beautiful [art] of today and the most praised [art] of ancient times should utterly perish by the negligence of modern men. Therefore, most illustrious and reverend Laelius, if you obtain any leisure among the great and most important occupations by which you are distracted, and if you sometimes retreat and withdraw to the study of those matters in which from youth you greatly delighted, do not spurn these. For in reading this book take, if not utility, a pleasure certainly noble and not unworthy of a liberal man.[58]

Baldi thus admits that this ancient treatise on how to make ancient weapons is neither useful nor practical; rather, he recommends it as an inspiration for modern inventors and as a liberal literary and historical diversion from practical affairs.

Nevertheless, in the Life of Hero that was printed with the *Belopoiica*, Baldi castigated Cardano for saying that although what is taught in the *Pneumatica* is ingenious and pleasant, it bears little application and utility to human uses.

> For who would say and impute [Baldi objects] that those things are useless that provide honest pleasure and are able to refresh with a certain sweet and harmless pleasure and lift up the soul tired with the weight of cares? or who would deny, unless rashly and stubbornly, that cupping glasses, lamps, syphons, clocks, and another 600 of such great instruments bring utility and convenience widely to human need for daily use?[59]

True, but were such amusing and useful things learned first from Hero?

If the *Belopoiica* was impractical because it described how to build obsolete weapons, the *Automata* and the *Pneumatica* were merely amusing and entertaining because they concerned largely frivolous though ingenious toys. And because the pneumatic and hydraulic devices described in the *Pneumatica* depended on principles different from those of the rest of mechanics, they were not easily assimilated to the rest of mechanics as it emerged as a mathematical science. Nothing in Hero's *Pneumatics* suggested that these principles—air and water pressure, and the vacuum – could be used to *multiply* power, which was the main purpose of all the other mechanical devices that Hero himself had described in his *Mechanics*. And so they were classified by their common uses as entertainments and magic-working, rather than brought under the common principles of the rest of mechanics. Galileo was perhaps the first to suggest that the principle governing the multiplication of power in all the other mechanical devices—that force and speed are inversely proportional—also applied to hydraulics, though it was Archimedes' hydrostatics rather than Hero's pneumatics that was his inspiration.[60] Galileo, however, was

---

[58]Hero of Alexandria, *Heronis Ctesibii Belopoeeca, hoc est. Telifactiva*, tr. Bernardino Baldi (Augsburg, 1616), fol. A2; thanks to Roland Jeffreys for help with this translation.

[59]Baldi, *Heronis vita*, in *Belopoeeca* , 73; a version of Baldi's "Life of Hero," in Italian, thought to be the autograph, is in Milan, Biblioteca Ambrosiana MS. D 332 inf., fols 103r–108v.

[60]Galileo, *Discourse on Bodies in Water*, theorem IV, corollary II, tr. Thomas Salusbury (London, 1663), rpt. in facsimile, ed. Stillman Drake (Urbana: University of Illinois Press, 1960), 17–18.

also among the first that took Hero's theory of matter and the interparticulate void seriously, applying it, in the first of his two new sciences, to explain the cohesion and strength of material bodies.[61] So while Hero's five powers or simple machines had exerted their influence, through Pappus, on the mechanics of the later sixteenth century, his pneumatics and his theory of matter and the void had to await the seventeenth century before their full effects were felt.

[61]Galileo, *Discorsi*, First Day, tr. Stillman Drake, *Discourses on Two New Sciences* (Madison: University of Wisconsin Press, 1974); see A. Mark Smith, "Galileo's Theory of Indivisibles: Revolution or Compromise?" *Journal of the History of Ideas*, 37 (1976): 571–588.

Chapter 9
Duytsche Mathematique and the Building of a New Society: Pursuits of Mathematics in the Seventeenth-Century Dutch Republic

Fokko Jan Dijksterhuis

Abstract In the seventeenth-century Dutch Republic mathematicians and mathematics acquired notable social and intellectual prestige. They contributed to the establishment of a new state, first through practical projects of fortification, navigation, land management, and later also through learned pursuits in academia and cultural circles. It can be said that the Republic provided particularly fertile grounds for academic pursuits, through its make-up of distributed wealth and power and its economic characteristics. The various towns and provinces provided various settings and opportunities to aspiring mathematicians. This chapter compares two notable sites, the provinces of Holland and Friesland, whose parallels and particularities put into perspective the interactions between mathematics and society in the Golden Age of the Dutch Republic.

9.1 Introduction

In its Golden Age, the Dutch Republic had a favourable climate for the pursuit of mathematics. Practitioners found employment with towns and provinces in the development of the new society, savants cultivated the metamorphosizing mathematical *scienze*, the cultural and political elite appropriated the new *esprit géométrique*. People engaged in mathematics were a motley company, ranging from arithmetic teachers like Willem Bartjens, to surveyors like Jacob van Wassenaer, from professors like Adriaan Metius, to 'amateurs' like Christiaan Huygens, statesmen like Johan de Witt, and so on. Mathematics in the early Dutch Republic was a multifaced enterprise that yielded a large variety of intellectual and material production. The pursuit of mathematics flourished on a marked interest of the social elite in things mathematical, because of its utilitarian as well as its cultural value.

F.J. Dijksterhuis (✉)
University of Twente, Enschede, The Netherlands
e-mail: f.j.dijksterhuis@utwente.nl

L.B. Cormack et al. (eds.), *Mathematical Practitioners and the Transformation of Natural Knowledge in Early Modern Europe*, Studies in History and Philosophy of Science 45, DOI 10.1007/978-3-319-49430-2_9

I will sketch how mathematicians and mathematics acquired social and intellectual prestige in the Dutch Republic in the seventeenth century. Two phases can be discerned: societally oriented, practical mathematics in the early seventeenth century, expanded towards scholarly inclined pursuits towards the middle of the century. At the start of the century mathematicians placed themselves in the service of the Stadholders and successfully acquired a central role in state building. Textual pursuits played a noticeably prominent part in this. The term 'Duytsche Mathematique' comes from the program of the engineering school established in Leiden in 1600 and denotes the teaching of mathematics theory in the vernacular. In the middle of the century some mathematicians distanced themselves from the practical context of surveying and fortification and seized opportunities to tie in with the elite's cultural interest in things mathematical. Through this route, mathematics became a contributing factor to the budding new philosophies of the seventeenth century. This development was historically tied to the Leiden 'Duytsche Mathematique', which therefore forms a natural focus for a discussion of the role of practical mathematics in the transformations of natural knowledge in the seventeenth-century Dutch Republic.

However, the 'Duytsche Mathematique' cannot be the sole focus, for it was a Holland affair. The Republic consisted of different provinces and there mathematics was pursued as well. The Republic was not a social and political unity.[1] Besides being an association of seven provinces (and several subordinate territories) and having its political power divided among several institutions, no less than two Stadholders led the revolting provinces. During the second half of the sixteenth century the provinces of the Low Countries had revolted against the Spanish rule to secure local priviliges and religious freedom.[2] The first stage of the revolt was led by William of Orange (1533–1584) from the Nassau house in the German empire, until he was assassinated by an anti-protestant militant. In 1584, Willem Lodewijk of Nassau (1560–1620) had become the first Frisian Stadholder. His nephew Count Maurits of Orange (1567–1625) became Stadholder in the Hague the next year.[3] Willem Lodewijk and Maurits had grown up together in Nassau and side by side they pursued the tasks of govermentally and militarily establishing and securing the new state. They were important innovators of warfare in which their particular interests largely complemented each other. The two Stadholderly courts of The Hague and Leeuwarden, and the respective universities in Leiden and Franeker, were two distinct social, political and cultural centers. The pursuit of mathematics in both centers displayed basic similarities as to the goals and values, but differed in

[1] J. Israel, *The Dutch Republic. Its Rise, Greatness, and Fall 1477–1806* (Oxford: Oxford University Press, 1998), 276–306.

[2] O. Mörke, *Wilhelm von Oranien (1533–1584). Fürst und "Vater" der Republik* (Stuttgart: Kohlhammer, 2007).

[3] Friesland and Groningen (and Drenthe) chose Willem Lodewijk; Holland, Utrecht, Gelderland, Overijssel and Zeeland Maurits.

its institutional and conceptual realization.[4] I present a twin-image of Holland and Friesland in order to show in a historically rich way how mathematics developed within Dutch culture.

9.2 Establishing Mathematics for a New Society

Willem Lodewijk tried to build up a 'modern' society in Friesland. In his state building, he stimulated the development of an intellectual life with two nuclei: Calvinist theology and practical mathematics.[5] The first was connected with the establishment of a Reformed society that had liberated itself from the Spanish king. The second was important because of the efforts the continuing war with Spain required from the Stadholders. In the formation of a strong community of faith and a powerful army, Willem Lodewijk also sought intellectual reinforcement. He had a marked interest in the classics, having studied with Lipsius in Leiden and he extensively read Roman military texts.[6] On this basis he introduced the volley technique, which in its turn fundamentally changed battle tactics.[7] Willem Lodewijk's penchant for scholarship was also seen at the Stadholderly court. He gathered scholars and ideas round him and organized 'Erasmian' tables: serious conversations over an abstemious meal where the emphasis was on concrete matters rather than lofty ornamentations.[8]

Willem Lodewijk not only saw to it that mathematics intellectually and practically furnished his new society, but also that it fashioned his own claims of sovereignty over this new society. The rector of the Groningen University, Ubbo Emmius (1547–1625) acted as Willem Lodewijk's chorographer. He wrote extensive geographies and histories of Friesland in which he established the geographical and historical identities of the Frisians and their Stadtholder emphasizing their ancient roots.[9]

In the 1580s Willem Lodewijk developed the school in Franeker into an official university that was formally established in 1585. The intellectual themes underlying his societal conception stood central: Calvinist theology and practical mathematics.

---

[4]See also, K. van Berkel, "Het onderwijs in de wiskunde in Franeker in vergelijkend perspectief," *It Beaken* 47 (1985):220–222.

[5]W. Bergsma, "Willem Lodewijk en het Leeuwarder hofleven," *It Beaken* 60 (1998):199–201 and 215–222. Israel, *Dutch Republic*, 569–572.

[6]Israel, *Dutch Republic*, 267–171. Ch. van den Heuvel, "Wisconstighe Ghedachtenissen. Maurits over de kunsten en wetenschappen in het werk van Stevin," in *Maurits, Prins van Oranje*, ed. K. Zandvliet (Zwolle: Waanders, 2000),113–116.

[7]G. Parker, *The Military Revolution. Military Innovation and the Rise of the West, 1500–1800* (Cambridge: Cambridge University Press, 1988), 18–20. The volley technique is the coordinated firing by a group of soldiers: the first row fires, steps to the back to reload, and lets the second fire. With some five rows a continuous firing is possible. The technique requires highly trained soldiers.

[8]Bergsma, "Willem Lodewijk en het Leeuwarder hofleven," 215–227.

[9]U. Emmius, *Guilhelmus Ludovicus Comes Nassovius* (Groningen: Johannes Sassius, 1621).

Adriaan Metius (1571–1635) was instrumental in giving shape to the latter pillar.[10] In 1598 Willem Lodewijk recruited Metius for the chair of mathematics in Franeker. He was the second son of Adriaan Anthonsz, the chief fortificationist of Maurits' and Willem Lodewijk's armies. He had studied in Franeker in 1589, switching to Leiden in 1594 to pursue his interest in mathematics. He studied with Rudolph Snellius (1546–1613), the father of Willebrord. Metius stayed with Tycho at the Hveen observatory for some time to be initiated in instrumental astronomy. Thereafter he gave private courses at the German universities Rostock, Marburg and Jena, before returning to the Republic where he assisted his father briefly.[11] In 1598, Willem Lodewijk advised him to register again in Franeker, holding out to him the prospect of a professorate in mathematics. The same year Metius was appointed extraordinary professor of mathematics, becoming full professor in 1600. On this occasion Metius received permission to lecture in both Latin and Dutch and to promote any candidate in mathematics. The permission to teach in the vernacular opened the possibility of educating engineers and surveyors, an activity that clearly met Willem Lodewijks aspirations. Metius' students of practical mathematics were not automatically licensed as practicing surveyors, though, they first had to be admitted by the 'Hof van Friesland'. After Metius' death in 1635, a surveyor school was institutionalized at Franeker University in 1641.[12]

In addition to his teaching activities Metius shaped his professorate, as well as his patronage relationship with Willem Lodewijk, in a range of textbooks. In these he explained established knowledge of practical mathematics and introduced recent theoretical and practical developments to his Frisian public. *Arithemeticae & Geometriae Practica* (1611/1625/1626) contained an exposition of surveying, in which Metius discussed the construction and operation of the measuring chain and the astrolabe and introduced the method of triangulation. Metius provided a basic network for the Frisian cities, apparently following the example of Willebrord Snellius' triangulation project in Holland.[13] He further treated Galileo's proportional compass and the 'Old-Dutch Fortification System'. Metius' textbooks provided – in variable degrees of abstraction – a scholarly rendering of practical affairs of

[10]One Johannes Roggius had preceded him, but he had only stayed for a short time and left after internal controversies at the university. The historical overview in this paragraph draws primarily on Berkel, "Onderwijs," 215–216.

[11]Arjen Dijkstra, *Between Academics and Idiots: A Cultural History of Mathematics in the Dutch Province of Friesland (1600–1700)*, Ph.D. thesis, Universiteit Twente, 2012. See also H. Terpstra, *Friesche Sterrekonst. Geschiedenis van de Friese sterrenkunde en aanverwante wetenschappen door de eeuwen heen* (Franeker: Wever, 1981) 55–59.

[12]P.J. van Winter, *Hoger beroepsonderwijs avant-la-lettre. Bemoeiingen met de vorming avn landmeters en ingenieurs bij de Nederlandse universiteiten van de 17^e^ en 18^e^ eeuw* (Verhandelingen der Koninklijke Nederlandse Akademie van Wetenschappen, Afd. Letterkunde, Nieuwe Reeks, deel 137) (Amsterdam: Noord-Hollandsche Uitg. Mij., 1988), 46–54.

[13]H.A.M. Snelders, "Alkmaarse natuurwetenschappers uit de 16de en 17de eeuw," in *Van Spaanse beleg tot Bataafse tijd. Alkmaars stedelijk leven in de 17de en 18de eeuw* (Alkmaarse historische reeks, 4) (Zutphen, 1980) 101–122.

navigation, surveying and the like. The publications in Latin tend to be more theoretical, whereas the publications in Dutch are more practically-oriented. For example, in *Manuale arithmeticae & geometricae practicae* (1633), a translation and adaptation in Dutch of the *Practica*, the theory of arithmetic and geometry is stripped down to its bare essentials, giving full emphasis to guidelines of reckoning, surveying and fortress-building. The *Manuale* also added an exposition of Napier's rods. In this way Metius introduced recent developments in practical mathematics in Friesland.

It will come to no surprise that Metius was quick to introduce the telescope to Friesland. He first discussed the instrument in his *Institutiones astronomicae & geographicae* (1614), a Dutch edition of *Institutiones astronomicarum* (1608). Metius' journalistic swiftness does not come as a surprise if we bear in mind that his brother Jacob was a builder of telescopes and some held him to be the inventor of this instrument.[14] In the *Institutiones*, Adriaan described telescopic observations made by his brother: sunspots, Jupiter's satellites and the stars of the Milky Way, and so on. He emphasized the novelty of these observations "which have been known to no authors, as being seen only by the distant views (telescopes) that have been found by my brother Jacob Adriaanz. about 6 years ago."[15] Jacob appears to have been a very secretive person who showed his instruments, in particular the later improved ones, to hardly anyone. The contrast with the natural communicator Adriaan can hardly have been more marked.[16]

Willem Lodewijk brought Metius to Friesland to cultivate mathematics for the benefit of the conduct of war and civic administration, an assignment Metius carried out dutifully by elaborating a body of practically-oriented knowledge that kept pace with recent developments of practical mathematics. In his teachings he introduced state-of-the-art practical mathematics to the new society. In the 1626 *Arithmetica* he explained that lands that did not have the natural resources to develop a good life, could nevertheless realize this by developing the arts of navigation and the like.[17] He acquired Tychonian instruments that established Franeker as a site of astronomical observation.[18] Metius did not just serve his patron, he also pursued his own career.

---

[14] A. van Helden, *The Invention of the Telescope* (Transactions of the American Philosophical Society held at Philadelphia for promoting Useful Knowledge. Volume 67, part 4) (Philadelphia, 1977), 5–6.

[15] A. Metius, *Institutiones Astronomicae et Geographicae* (Franeker, 1614), 3: "dewelcke by ghene Autoren zijn bekent gheweest, dan werden alleene ghesien door de verre ghesichten, die by mijn Broeder Jacob Adriaenz. over omtrent 6 jaren ghevonden zijn geweest."

[16] Recently, Huib Zuidervaart has mapped the life and work of Jacob Metius in much detail, using new sources and qualifying older claims considerably. H. Zuidervaart, "The 'Invisible Technician' Made Visible: Telescope making in the Seventeenth and early Eighteenth-century Dutch Republic," in *From Earth-bound to Satelite. Telescopes, Skills and Networks*, ed. G. Strano, et al. (Leiden: Brill, 2011): 41–102.

[17] A. Metius, *Arithmeticae libri duo et Geometriae* (Leiden, 1626), 124.

[18] A. Dijkstra, "A Wonderful Little Book. The Dissertatio Astronomica by Johannes Phocylides Holwarda (1618–1651)," in *Centres and Cycles of Accumulation in and around the Netherlands in the Early Modern Period*, ed. L. Roberts (Berlin: Lit, 2011): 73–100.

His textbooks informed the new Frisian state as well as they fashioned his academic ambitions. Metius became a well-known mathematician and a respected educator, attracting students from all over Europe. When Descartes came to the Low Countries in 1629, he first settled in Franeker and kept company with Metius.[19]

As a mathematician Metius attracted the attention of statesmen and acquired a key role in the building of the new state. This explicitly included theoretical aspects of mathematics aimed at reinforcing mathematical practice, in the same way academic theology would reinforce Calvinist preaching. The opportunity to do so was created by the changes in warfare brought about by the specific nature of the Dutch fight for independence. The defense system was characterized by a tight network of fortifications and fixed garrisons that called for pervasive engineering and a high degree of discipline.[20] The textual bias of army organization can be seen in the use of illustrated instructions to implement standardized drilling throughout the ranks.[21]

9.3 Establishing Duytsche Mathematique

The Holland counterpart of Metius was Simon Stevin (1648–1620), who established a prominent role for mathematics through his relationship with Count Maurits. The pairs Stevin–Maurits and Metius–Willem Lodewijk had similar ambitions regarding the use of mathematics in statebuilding. However, their relationships differed and the realization of mathematics initiatives in Holland and Friesland differed accordingly. In the first place, Stevin was not at a university, and the teaching of practical mathematics would be organized within a separate institution. Secondly, Stevin was more directly involved in military affairs and fortification in particular. Lastly, his relationship with Maurits was more personal and they collaborated directly on mathematical topics.[22] The contact between Stevin and Maurits probably went back to the early 1580s when they both studied in Leiden. Maurits was directly interested in mathematics and even made some original contributions.[23] Willem Lodewijk's main interest was classical military reading and he left mathematics to Metius. Furthermore, Willem Lodewijk always sought practical applications of his readings, whereas Maurits had a prediliction for theoretical experiments and

[19] W.R. Shea, *The Magic of Numbers and Motion. The Scientific Career of René Descartes* (Canton, Mass.: Science History Publications USA, 1991), 191.

[20] F. Westra, *Nederlandse ingenieurs en de fortificatiewerken in het eerste tijdperk van de Tachtigjarige Oorlog, 1573–1604* (Canaletto: Alphen aan de Rijn, 1992), chapters 7, 9 and 11 in particular.

[21] Parker, *Military Revolution*, 18–23.

[22] Heuvel, "Wisconstighe Ghedachtenissen," 107–108.

[23] *Ibid.*, 108–110.

elaboration.[24] In 1593 Stevin formally entered Maurits' service as an engineer as well as personal teacher. Their intellectual exchange was embodied in Stevin's *Wisconstighe Ghedachtenissen* (*Mathematical Thoughts*, 1605–1608).

From the perspective of mathematics, the collaboration between Stevin and Maurits was crowned by the establishment in 1600 of an engineering school in Leiden.

> As it has pleased His Excellency, Count Maurits of Nassau, Stadholder of Holland, and Captain General, that, for the benefit of the state, here in the university should be taught in good Dutch language the art of counting and surveying, principally for the advancement of those who should want to become engineer...[25]

Although it was connected to the university, the engineering school was a separate institution. The chair of practical mathematics was new and existed independently of the university chair of mathematics, held by Rudolph Snellius at that time. In contrast, in Franeker instruction in practical mathematics was given at the university, by the professor of mathematics, Metius. Stevin drew up the curriculum for the instruction in practical mathematics, but he would not carry it out. Ludolf van Ceulen and Simon van der Merwen became the first professors.[26] The reasons for establishing a separate institution rather than assigning the teaching of mathematics in the vernacular to the professor are complex and are yet to be investigated in detail. On the one hand, Maurits may have had a kind of *Ritterschule* in mind as they existed at many German courts. On the other hand, the climate at the university had recently turned rather against practical pursuits after the strong humanist direction advocated by Joseph Scaliger (1540–1609) had become dominant.[27]

Maurits and Stevin called the engineering training the 'Duytsche Mathematique'. Stevin wrote the program that accompanied Maurits's request to the university curators. It was to teach surveyors and fortificationers a body of mathematics theory in Dutch concentrating on practically relevant topics. It deserves notice that it was not self-evident that fortificationists would be taught mathematics theory, rather than be trained in the field.[28] The establishment of the 'Duytsche Mathematique' bears

---

[24]*Ibid*., 117–119.

[25]P. Molhuysen, *Bronnen tot de geschiedenis der Leidsche Universiteit*. Vol. 1 (Rijksgeschiedkundige Publicatiën 20) (Den Haag, 1913) 122. "Alsoo Sijne Excellentie, Grave Maurits van Nassau, Stadthouder van Hollant, ende Capiteyn Generael, tot dienst van den lande goetgevonden hadde, dat in de Universiteit alhyer soude worden gedoceert in goeder duytscer tale die telconste ende lantmeten principalycken tot bevordering van de geenen die hen souden willen begeven tottet ingenieurscap ...."

[26]Winter, *Hoger beroepsonderwijs*, 14–16.

[27]H. Hotson, *Commonplace Learning. Ramism and its German Ramifications, 1543–1630* (Oxford: Oxford University Press, 2007).

[28]J.A. Bennett, "The Challenge of Practical Mathematics," in *Science, Culture and Popular Belief in Renaissance Europe*, ed. Stephen Pumfrey et al. (Manchester: University of Manchester Press, 1991),180–182. E. Taverne, *In 't land van belofte: in de nieue stadt. Ideaal en werkelijkheid van de stadsuitleg in de Republiek. 1580–1680* (Maarssen: Schwartz, 1978), 49–81.

the mark of Stevin's particular conception of the pursuit of mathematics, aimed at integrating '*Spiegheling*' (contemplation) and '*Daet*' (action).[29] This combination of theory and practice was the heart of Stevin's program of the 'Duytsche Mathematique'. Stevin's curriculum prescribed in detail what mathematics the instructors should teach.

> To this end one will teach arithmetic or counting and surveying but only so much of each as is required for practical, common engineering.[30]

For example, regarding the determination of areas Stevin stipulated:

> The measuring of circles with segments of that sort, further the area of spheres. The shapes named ellipsis, parabola, hyperbola and the like, that is not necessary here, because engineers are very seldom made to perform such measurements; but only they shall learn with straight planes, after that curvilinear in surveyor's manner, measuring thus a plane by various division, like in triangles or other planes to see how this matches with that.[31]

Despite the different ways in which the pursuit of mathematics was organized in Holland and Friesland, the 'Duytsche Mathematique' reflected conceptions of useful knowledge similar to those of Willem Lodewijk and Metius. The new Republic, in the middle of liberating itself from Spanish rule, did not just need skillful hands, but hands that were also informed by learning. Action with contemplation, as Stevin said.

The alliance between Holland and Friesland was illustrated by two books on surveying published in the same year the 'Duytsche Mathematique' was established. *Practijck des Lantmetens* (*Practice of Surveying*, 1600) and *Van het gebruyck der geometrische instrumenten* (On the Use of Geometrical Instruments, 1600) were published by the Jan Pieterszoon Dou from Holland and Johan Sems from Friesland together. They expounded similar conceptions about theory and practice in surveying as Stevin and Maurits held. Their books sold well and were standard repertoire for surveyors, but did not realize their aim at establishing an official training for surveyors. The 'Duytsche Mathematique' was a training for military engineers and would not provide formal qualifications for surveying.

The 'Duytsche Mathematique' nicely illustrates the close tie between mathematics, discipline and defense. The first professor was Ludolf van Ceulen (1540–1610). The lessons would be given in the Faliebegijnkerk, where the university library and

---

[29]K. van Berkel, "The Legacy of Stevin. A Chronological Narrative" in *A History of Science in the Netherlands. Survey, Themes and Reference*. ed. Klaas van Berkel, Albert van Helden, Lodewijk Palm (Leiden: Brill, 1999), 16–20.

[30]Molhuysen, *Bronnen*, 389*: "Hyer toe sal men leeren die arithmeticque oft het tellen ende het landtmeten maer alleenlyck van elck soe veel, als tottet dadelyck gemeene ingenieurscap nodich is."

[31]Molhuysen, *Bronnen*, 390*: "Het meten des rondts mette gedeelten van dien aengaende, voerts het vlack des cloots, de formen genaemt ellipsis, parabola, hyperbole ende diergelijcke, dat en is hyer nyet nodich, wantet den ingenieurs seer selden te voeren compt, sulcke metinge te moeten doen; maer alleenlyck sullense leeren met rechtlinige platten, daer na cromlinige landtmetersche wijse, metende alsoe een plat deur versceyde verdeelinge, als in dryehoucken of ander platten om te syen hoe t'een besluyt met het ander overcompt."

anatomical theater were already located. In the room under the library Van Ceulen had been giving fencing lessons since 1594. Van Ceulen was succeeded in 1615 by Frans Van Schooten Sr., who established a tradition in 'Duytsche Mathematique' that would sustain well into the century. The backbone of the program was the so-called Old-Dutch Fortification System, as it had developed under Maurits and had been codified by Stevin in *Sterctenbouwing* (*Stronghold construction*, 1594). In the winter van Schooten taught the theory of fortification, in the summer he attended field practice with the army.[32] Van Schooten Sr. started somewhat of a dynasty at the engineering school, being succeeded in 1645 by his son Frans Jr. who in his turn was succeeded by his half-brother Petrus in 1660, continuing the tradition of 'Duytsche Mathematique' until the 1670s.

9.4 Cultivating Mathematics for a New Philosophy

Although dutifully serving as professor of Duytsche Mathematique, Frans van Schooten Jr. (1615–1660) looked for new routes to realize the cultural capital of mathematics. Dutch society was changing by that time. The Revolt had been successful and although the war continued until 1648, the immediate threat had diminished. The focus of building work shifted from siege and fortification to land reclamation and city extension, altering the demand for mathematical skills. A civic society developed in which a patrician elite increasingly established a firm position and began acting like a new aristocracy. Van Schooten used his, and his family's, position as a stepping stone to move upward socially and culturally in this new society. He gave the Duytsche Mathematique a new twist, distancing it from the practical mathematics of his father and seeking alliance with the interests of the elite. Van Schooten studied at Leiden University, with Jacobus Golius (1596–1667), professor of Arabic and successor of Snellius at the chair of mathematics.[33] He started replacing his father at the Engineering School in 1635 until he succeeded him in 1645.[34] In the intervening years he had established relations within the Dutch elite and with prominent French mathematicians. He acquainted himself with the new mathematics of Descartes, Viète and Fermat. Or rather geometry, as the word mathematics was used in the seventeenth century for the less lofty practices of measuring and calculating.[35]

---

[32] Taverne, *In 't land van belofte*, 64–66.

[33] F. Dijksterhuis, "Moving Around the Ellipse. Conic Sections in Leiden (1620–1660)," in *Silent Messengers. The Circulation of Material Objects of Knowledge in the Early Modern Low Countries*, ed. Sven Dupré and Christoph Lüthy (Berlin: Lit, 2011): 89–124.

[34] J. Hofmann, *Frans van Schooten der Jüngere* (Wiesbaden: Steiner, 1962), 1–2.

[35] Compare Olmsted, John W., "Jean Picard's 'Membership; in the Académie Royale des Sciences, 1666–1667: the Problem and its Implications," in *Jean Picard et les Débuts de l'Astronomie de Précision au XVIIe Siècle*, ed. Guy Picolet (Paris: Édition du Centre National de la Recherche Scientifique, 1987), 85–116.

With Golius Van Schooten first met Descartes, who had come to Leiden in 1630. He quickly became one of Descartes' favorites and assisted him on several projects. He made the illustrations for the essays of *Discours de la Methode* and drew a template of a hyperbola for the grinding of a non-spherical lens. This latter project was organized by Constantijn Huygens, who was introduced by Golius to Descartes in 1635.[36] To Van Schooten the participation of Huygens meant a direct access to the Holland elite. Huygens was a prominent figure in the highest political and cultural ranks; he was secretary to the Stadholder, a renowned poet and composer, and the principal cultural intermediary in the middle of the seventeenth century.

The association with Golius created opportunities for Van Schooten to go beyond the milieu of the Duytsche Mathematique. Around 1639 he wrote an introduction to Descartes' geometry, a basic exposition of the new method of letter calculation.[37] Sending it to Mersenne, Van Schooten used it as his introduction to the Republic of Letters. He later published it as *Principia Matheseos Universalis* (1651). Around the same time he struck a deal with the Leiden publisher Elzevier to collect writings of the new French mathematicians. He traveled to France in 1641, where he copied several manuscripts of Fermat and Viète. It resulted in the publication of *Francisci Vietae Opera mathematica* with Elzevier in 1646.[38] The same year Van Schooten had published his first original work, *De organica conicarum sectioneum in plano descriptione*. It was an exposition of the kinematic generation of conic sections that combined artisanal and academic facets of mathematics. On the one hand it treated the practical drawing of ellipses, hyperbolas and parabolas, proposing new instruments useful for gardeners, architects and the like. On the other hand it elaborated the mathematical foundations and consequences of the procedures proposed, much in the way Mydorge and Descartes did, by embedding it in the classical theory of Apollonius. Thus *Organica* constituted a crossroads between the 'Duytsche Mathematique' of his father, the classical *geometria* of Golius, and the new *géométrie* of Descartes and Viète.[39]

Van Schooten was part of an extended circle of mathematicians courting the Dutch elite. All kinds of mathematicians competed over positions as teachers, advisors, examiners. To succeed his father in 1645, Van Schooten had to compete with Jan Stampioen (1610–?1689), the mathematics tutor of Constantijn Huygens's sons. When Van Schooten got the position, Stampioen sought revenge by securing a position as provincial examiner of surveyors who would judge the competences of

[36] W. Ploeg, *Constantijn Huygens en de Natuurwetensc* happen (Rotterdam: Nijgh & Van Ditmar, 1934), 36–38. F. Dijksterhuis, "Constructive Thinking. A Case for Dioptrics", in *The Mindful Hand. Inquiry and invention from the late Renaissance to early industrialisation*, ed. L. Roberts et al. (Amsterdam, 2007), 59–82.

[37] F. van Schooten, "Calcul de Mons. Des Cartes," in: Descartes, René, *Oeuvres de Descartes*, ed. Charles Adam and Paul Tannery, 2nd edn., 11 vols. (Paris: 1974–1986), vol 10, 659–680.

[38] Hofmann, *Frans van Schooten der Jüngere*, 2–3.

[39] Dijksterhuis, "Moving Around the Ellipse," 106–107.

Van Schooten's students.[40] However, Van Schooten's mathematical pursuits went well beyond the original 'Duytsche Mathematique', leading abroad to the new geometry in France, using Latin rather than Dutch. He gathered the writings (and acquaintance) of prominent mathematicians and rendered them into a didactically appropriate form. The acme of Van Schooten's oeuvre would become his adaptation and translation into Latin of Descartes' *La Géométrie*.

The ambivalence between Van Schooten's official position as professor of 'Duytsche Mathematique' and his geometrical work was noticed by contemporaries as well:

> And in this church, where the English preach nowadays, in this beguinage, all days (except Wednesday and Saturday) from 11 to 12 o'clock, public lessons are given in the Dutch language, on the mathematical arts, for the convenience of the unlettered, like bricklayers, carpenters, and the like; who at that time find themselves here in crowds without coats but equipped with their sticks, aprons, etcetera; which then is very farcical to see. The professor, who gives Dutch lessons, nonetheless in his usual distinguished professor gown, or coat, (like al the other Latin professors do theirs,) is the very learned, and widely renowned sir Franciscus van Schooten.[41]

With his work in the new geometry Van Schooten developed extra cultural capital that extended beyond elementary mathematics, appealing to the intellectual interests of the patrician elite. In the 1650s, The professor of 'Duytsche Mathematique' began attracting a new kind of students: patrician sons aiming at an academic education rather than professional training. Van Schooten had acquired enough status to have the young Huygenses, the young De Witt, the young Hudde, the young Heuraet come and study with him. Why did they not go to the real professor, instead of this teacher of the masses?

The patriciate's ties with the Duytsche Mathematique are historically rooted in the early phase of the Dutch Republic. Yet, it increasingly distanced itself from the common businesses of navigation, surveying and fortification, turning themselves in 'nouveau' aristocrats with matching intellectual interests towards

---

[40]F. Dijksterhuis, "Stampioen Jr., Jan Janszoon (1610–after1689)," in *The Dictionary of Seventeenth and Eighteenth-Century Dutch Philosophers*. 2. vols., ed. W. van Bunge et al. (Bristol, 2003), 938–940. F. Dijksterhuis, "Fit to Measure. 'Bequamheit' in Mathematics in the Dutch Republic," in *Public Offices, Personal Demands. Capability in Governance in the Seventeenth-Century Dutch Republic*, ed. J. Hartman and J. Nieuwstraten eds. (Newcastle: Cambridge Scholars Publishing, 2009), 80–100. Van Schooten had also decided against Stampioen in the latter's controversy with Descartes.

[41]J.N. Parival, *De Vermaecklijckheden van Hollandt* (Amsterdam, 1660), 188–189: "En in die Kercke, waer de *Engelsche* nu predicken, in dit *Bagijne-Hoff*, worden alle dagen, (behalven 's Woensdaeghs, en Saterdaeghs) van elf tot twaelf uuren, openbare Lessen gedaen in de Neerlandsche Tael, in de *Mathematische* Konsten, tot gerief van de ongeletterden, als *Metselaers*, *Timmer-luyden*, en diergelijcke meer ; die haer dan met hoopen in die tijdt hier vinden : sonder mantels, maer met hare stocken, en schoots-vellen, &c. versien ; dat dan seer kluchtigh om sien is. Den *Professor*, die duytsche lessen voor haer doet, evenwel in sijnen gewoonlijcken aensienlijcken *Professor*-Tabbaert, ofte Rock, (soo wel als alle de andere *Latijnsche Professoren* de hare doen,) is den Hoogh-geleerden, en Wijdt-vermaerden D: Franciscus van Schooten."

the middle of the century.[42] For the new generations of patricians, mathematics became a cultural capital that went beyond its practical value. In the pursuits of De Witt, Hudde and Huygens we can see a particular mathematical ideology. They regarded mathematics as a model of rationality and a source of lucid thinking crucial to general education.[43] This conception of mathematics had its roots in the Renaissance.[44] Constantijn Huygens, Christiaan's father, was heavily influenced by Renaissance ideas.[45] The Dutch patricians studying with the professor of Duytsche Mathematique, however, had moved beyond a Renaissance notion of rational rhetoric to one that can be seen as an early instance of Enlightenment thinking, whereby reason steered by mathematics was the foundation of knowledge and judgment. I have the impression that, in the midst of the political and religious frictions that characterized the Dutch Republic in the seventeenth century, mathematics offered an intellectual haven to its future dignitaries.[46] In the meantime, a civic version of the 'Duytsche Mathematique' did not come into being. Despite the vast demand for mathematical skills in the large infrastructural projects of land reclamation and city extensions, no formal institution to train civil engineers was established by the patrician administrators.

We may say that Van Schooten had kept pace with this development and that his mathematics perfectly fitted the new inclinations of the patriciate. It was rooted in the Duytsche Mathematique but had outgrown it to become a new geometry of a more aristocratic stature. The result was Van Schooten's extended second edition of *Geometria à Renato Des Cartes* (1659–1661), which contained numerous contributions of his patrician pupils.[47] The *Geometria* constituted a further step beyond the 'Duytsche Mathematique' in comparison to the *Organica* of 1646. It was purely speculative mathematics, not oriented to practical issues of curve drawing (not to mention fortification).[48] In addition it pointed towards the new *physico-*

[42]L. Kooijmans, "Patriciaat en aristocratisering in Holland tijdens de zeventiende en achttiende eeuw," in *De Bloem der Natie. Adel en patriciaat in de Noordelijke Nederlanden*, ed. J. Aalbers (Meppel: Bloom, 1987), 98–103.

[43]Berkel, "The Legacy of Stevin," 52–59. On the role of mathematics in the education of 'honnêtes hommes' see M. Jones, *The Good Life in the Scientific Revolution. Descartes, Pascal, Leibniz, and the Cultivation of Virtue* (Chicago: University of Chicago Press, 2006).

[44]P.L. Rose, *The Italian Renaissance of Mathematics. Studies on Humanists and Mathematicians from Petrarch to Galileo* (Genève: Droz, 1975).

[45]F.J. Dijksterhuis, "Vader en Zoon. Over Constantijn en Christiaan Huygens," *Bzzlletin* 28 (1999): 18–22.

[46]Later in the seventeenth-century the mathematical approach in philosophy was criticized because of the association with Spinozism. The Newton-inspired 'physico-theology' provided an answer for the enlightened enthusiasts. See R. Vermij, "The formation of the Newtonian philosophy: the case of the Amsterdam mathematical amateurs," *The British Journal for the History of Science* 36 (2003): 183–200.

[47]Berkel, "The Legacy of Stevin," 54; Dijksterhuis, "Moving Around the Ellipse".

[48]Until the eighteenth century two dimensions were distinguished in the stratification of mathematics: subject matter and goal. Regarding the subject matter pure mathematics was contrasted with mixed, signifying the abstractedness of mathematical entities. Regarding the goals of mathematics,

mathematics of motions, light and the like that Descartes discussed in the other essays of *Discours de la Methode*.

The collaboration of Van Schooten and his patrician pupils for *Geometria* was the basis for the further development of 'aristocratic' mathematics during the second half of the seventeenth century. Whereas De Witt and Hudde focussed on their administrative duties and kept their mathematics private, Christiaan Huygens steered clear of the diplomatic career his father had in mind for him and devoted his life to the sciences. He transformed Van Schooten's teachings into a new *physico-mathematica* exemplified in his *Horologium Oscillatorium* of 1673. In the development of Huygens' optical studies between 1650 and 1680 the transition can be traced from the mathematics of lenses and telescopes to the mathematization of the mechanistic nature of light. Elsewhere I have argued that Huygens' wave theory historically was an extension of his dioptrics, transferring the concepts and techniques of the mathematical study of rays and instruments to the realm of unobservable waves.[49] Rather than developing Descartes' natural philosophical program of mechanizing nature, Huygens extended mixed mathematics into new domains developing a particular kind of mathematico-philosophizing. His work with Van Schooten on the *Geometria* had formed the starting point of Huygens' mathematics, the *Geometria* in its turn being the product of Van Schooten's development as mathematician and his successful establishment of relationships with the Dutch elite. In retrospect, we see how new ways of philosophizing were rooted socially and culturally in the 'Duytsche Mathematique'.

9.5 Back to Friesland

To conclude, I give a brief sketch of the developments that took place in the meantime in Friesland. Franeker university had been established to furnish the two pillars of Calvinist theology and practical mathematics with intellectual underpinnings. In 1652 the *Friesche Sterre-konst* (*Frisian Astronomy*) of the Franeker professor of logic Johannes Phylocides Holwarda (1618–1651) was published, which can be regarded as the synthesis of Willem Lodewijk's vision. Holwarda elaborated astronomy into a Calvinist metaphysical scheme.[50] As a student he had used Metius' instruments and discovered a new celestial phenomenon, nowadays known as the variable star Mira Cetis.[51] With Metius' successor at the chair of mathematics,

the practical mathematics was contrasted to the speculative. See H.M. Mulder, "Pure, Mixed and Applied Mathematics: The Changing Perception of Mathematics Through History," *Nieuw Archief voor Wiskunde* 1990, 4–8: 27–41.

[49] F.J. Dijksterhuis, *Lenses and Waves. Christiaan Huygens and the Mathematical Science of Optics in the Seventeenth Century* (Dordrecht: Springer, 2004), 225–235.

[50] Terpstra, *Friesche Sterrekonst*, 65–74.

[51] Dijkstra, "A Wonderful Little Book".

Bernard Fullenius sr. (1602–1657), he performed the astronomical observations of the *Friesche Sterre-konst*. In Holland, as we have seen, mathematics had been joined with the new philosophy of the day. In Friesland a similar course was taken towards contemplative pursuits, but here mathematics was connected to theology. The successor of Fullenius, Abraham de Grau (1632–1683) tried to combine the spectrum of philosophy into the currently developing 'historica philosophia' and set great store on mathematics.

The link with Huygens's *physico-mathematics* was established by Bernard Fullenius, jr. (1640–1707), who took his father's chair in 1684. Fullenius jr. was a Franeker patrician, comparable to Hudde and De Witt. However, in a move unthinkable for his Holland counterparts, he gave up his position as urban magistrate and became professor at the university. As professor of mathematics he established a network of savant exchange extending throughout the Republic. The nexus was formed by the secretary of the Frisian Stadholder, Philip Ernst Vegilin van Claerbergen, who introduced Fullenius in the 1680s to, among others, Christiaan Huygens.[52] Huygens found a kindred spirit, for Fullenius turned out to be well-versed in matters dioptrical, and in his will asked him to publish his posthumous papers.

The next phase concerns the development in the early eighteenth-century of the a natural philosophy founded upon mathematical principles and in which instruments stood central.[53] The nucleus were the informal societies that developed in the Holland cities in particular, but Friesland joined in in an interesting way. The pivot of early eighteenth-century mathematical culture in Friesland was Willem Loré (1679–1744), a protégé of Fullenius Jr. Loré was a man of humble origins who worked his way up by studying surveying in Franeker. He became lector under Fullenius teaching mathematics and surveying and government surveyor in 1707. Loré was the teacher of Wytze Foppes (1707–1778) and Jan Pietersz. van der Bildt (1709–1791) who started a line of Frisian telescope makers that continued through the entire eighteenth century. They too were of humble origins, originally being carpenters. Later members of this tradition also had their roots in the crafts, like the famous planetarium builder Eise Eisinga.[54] Besides having taught both carpenters, Loré played a stimulating role in their development as instrument makers and provided access for them and their products to the Stadholderly court. He became main assistant for the budding interest in the new philosophy at the Stadtholderly court. The then Stadholder, Willem IV, and his successor Willem V were highly interested

[52] A.F.B. Dijkstra, *Het vinden van Oost en West* (M.A.-thesis Groningen, 2007).

[53] Berkel, "The Legacy of Stevin," 68–76.

[54] H.J. Zuidervaart, Speculatie, *Wetenschap en Vernuft. Fysica en astronomie volgens Wytze Foppes Dongjuma (1707–1778), instrumentmaker te Leeuwarden* (Leeuwarden: Fryske Akad., 1995), 21–25; see also H.J.Zuidervaart, "Reflecting 'Popular Culture': The Introduction, Diffusion, and Construction of the Reflecting Telescope in the Netherlands," *Annals of Science* 61 (2004): 407–452.

in the sciences and instruments in particular.[55] They facilitated the creation of a physical theater at the Franeker Academy and of the position of an assistant. At the court the princes held scientific salons and built up a collection of instruments. The collection (and the Franeker demonstrator) went to The Hague in 1748. The previous year Willem IV had become Stadholder of the whole Republic and he moved his court to The Hague.

9.6 Conclusion

With the move of Willem IV to The Hague the two original courts of Willem Lodewijk and Maurits were united. This brings my sketch of the development of mathematics in the Republic to a close. Loré symbolyzes the reunion of Holland and Frisian branches like the surveying books of Sems and Dou had stood for the alliance a century earlier. Mathematical practices evolved alongside societal developments and I have argued how mathematicians tried to capitalize on the interests of the ruling elite. In the early days of the Republic two prominent sites for this process were established in the form of the Stadholderly courts of Leeuwarden and The Hague and their universities in Franeker and Leiden. I have expressly followed the Holland and Frisian branches seperately to show how societal setting and mathematical practice co-evolved. Both branches followed quite similar courses as regards the mathematical subject matter and orientation. At first the primary focus was on state-building and practices of fortification, surveying and so on. Later on, more academic practices were added, reflecting the aristocraticizing tendencies of the Dutch elites. However, the societal settings of Friesland and Holland differed and this is reflected in differences in the implementation of ideals regarding mathematics and the institutionalization of mathematical practices. So, in Holland an autonomous engineering school was established for instruction of practical mathematics, which in Friesland was embedded within the university. In Holland 'aristocratic math' became the processing of the new, French geometry, whereas in Friesland an amalgam of mathematics and (Calvinist) theology arose. Holland and Friesland did not, of course, develop seperately and in mathematics too, much interchange took place. Letters were sent, men of letters and of numbers travelled, and so on, an aspect that I have not discussed in any detail for this occasion. The co-evolution of Holland and Frisian mathematical cultures will be matter for further study.

Acknowledgements I would like to thank Arjen Dijkstra and Tim Nicolaije for their valuable comments and suggestions. This article is part of the NWO-funded research project "The Uses of Mathematics in the Dutch Republic" (016.074.330).

[55] P. de Clercq, "Science at Court: The Eighteenth-Century Cabinet of Scientific Instruments and Models of the Dutch Stadholders," *Annals of Science* 45 (1988): 113–152.

Combined Bibliography

Abbreviations

DNB *Oxford Dictionary of National Biography*, ed. by H.C.G. Matthew and Brian Harrison, 3rd ed. (Oxford: Oxford University Press, 2004); online edition edited by Lawrence Goldman, 2006–2008.

DSB *Dictionary of Scientific Biography,* ed. by Charles Coulston Gillispie (New York: Scribner, 1970–1990).

Primary Sources

Agricola, Georgius. 1556. *De Re Metallica*. Basel: H. Frobenium et N. Episcopium.

———. 1950. *De Re Metallica.* ed. and trans. Herbert and Lou Henry Hoover. New York: Dover Publications.

Alberti, Leon Battista. 1485. *De Re Aedificatoria (1452)*. Florence: Nicolai Laurentii.

———. 1988. *On the Art of Building in Ten Books*. Cambridge, MA: MIT Press.

Apianus, Peter. 1533a. *Cosmographicum liber*. Antwerp: Arnoldum Birckman.

———. 1533b. *Instrument Buch*. Ingolstadt: P. Apianus.

Aristotle. 1936. *Minor Works*. ed. and trans. Walter Stanley Hett. Cambridge, MA: Harvard University Press.

Ausonio, Ettore. 1968. Theorica speculi concavi sphaerici. In *Le Opere di Galileo Galilei*, ed. Galileo Galilei, Vol. volume 3. Florence: Speculi Concavi Sphaerici.

Bacon, Francis. 1627. *The New Atlantis*. London: J.H. for William Lee.

Bacon, Roger. 1659. Epistola de secretis operibus artis et naturae ed de nullitate magiae. In *Frier Bacon his discovery of the miracles of art, nature, and magick. Faithfully translated out of Dr Dees own copy, by T. M. and never before in English.* London.

Baldi, Bernardino. 1616. *Heronis Ctesibii Belopoeeca, hoc est Telifactiva*. Augsburg: Davidis Franci.

Barlow, William. 1597. *The Navigator's Supply Conteining many things of Principall importance*. London: G. Bishop, R. Newberry and R. Barker.

L.B. Cormack et al. (eds.), *Mathematical Practitioners and the Transformation of Natural Knowledge in Early Modern Europe*, Studies in History and Philosophy of Science 45, DOI 10.1007/978-3-319-49430-2

Beeckman, Isaac. 1939–1953. In *Journal tenu par Isaac Beeckman de 1604 à 1634*, ed. Cornelis de Waard, Vol. 4 vols. La Haye: M. Nijhoff.
Besson, Jacques. 1559. *De absoluta ratione extrahendi olea & aquas e medicamentis simplicibus*. Tiguri [Zurich]: Andream Gesnerum Jr..
———. 1569a. *L'Art et science de trouver les eaux et fontaines cachées soubs terre, autrement que par les moyens vulgaires des agriculteurs & architects*. Orleans: E. Gibier.
———. 1569b. *Instrumentorum et Machinarum. Liber Primus*. Orleans [or Lyons?].
———. 1578. *Theatrum instrumentorum et machinarum*. Leiden: B. Vincentium.
Birch, Thomas. 1760. *Life of Henry, Prince of Wales, Eldest Son of King James I*. London.
Biringucci, Oreste Vannocci. 1582. *Parafrasi di monsignor Alessandro Piccolomini sopra le Mecaniche d'Aristotile*. Rome: F. Zanetti.
Blagrave, John. 1585. *The Mathematical Jewel*. London: Walter Venge.
Bourne, William. 1578a. *The Arte of Shooting in Greate Ordenaunce*. London, [H. Bynnemann?].
———. 1578b. *Inventions or devices: Very necessary for all generalles and captaines, or leaders of men, as well by sea as by land*. London.
———. 1578c. *Treasure for Travellers*. London.
———. 1581. *An Almanacke and Prognostication for x. yeeres*. London.
———. 1587. *The Arte of Shooting in Great Ordnaunce*. London: [Thomas Dawson] for Thomas Woodcocke.
Busca, Gabriello. 1601. *Della Architettura Militare. Primo libro*. Milan: Bordone and Locarno.
Cataneo, Girolamo. 1564. *Opera nuovo di fortificare*. Bresica: Battista Bozola.
Corneweyle, Robert. 1972. *The Maner of Fortification of Cities, Townes, Castelles and Other Places, 1559*. Surrey: Richmond.
Davis, John. 1595. *The Worldes Hydrographical Discription*. London: Thomas Dawson.
de Bry, Theodore. 1590. *America. Pars I*. Frankfurt: Ioannis Wecheli.
de Caus, Salomon. 1612. *La perspective avec la raison des ombres et miroirs*. London.
de Parival, Jean Nicholas. 1660. *De Vermaecklijckheden van Hollandt*. Amsterdam: Samuel Imbrechts.
Dee, John. 1842. In *The Private Diary of Dr. John Dee*, ed. J.O. Halliwell. London: John Bowyer Nichols and Sons.
———. 1975. *The Mathematicall Praeface to the Elements of Geometrie of Euclid of Megara (1570)*. Introd. by Allen G. Debus. New York: Science History Publications.
del Monte, Guido Ubaldo. 1577. *Mechanicorum Liber*. Pesaro: Hieronymum Concordiam.
Descartes, René. 1964–1986. *Oeuvres de Descartes*, eds. Charles Adam and Paul Tannery. 2nd edn., 11 vols. Paris.
———. 1998. *Descartes, The World and Other Writings*. Trans. Stephen Gaukroger. Cambridge: Cambridge University Press.
Digges, Thomas. 1573. *Alaae seu scalae mathematicae*. London: Thomas Marsh.
Digges, Thomas. 1590. *An Arithmeticall Warlike Treatise Named Stratioticos*. London: Richard Field.
Digges, Leonard, and Thomas. 1571. *A Geometrical Practise, Named Pantometria*. London: Henrie Bynneman.
———. 1579. *An Arithmeticall Militare Treatise, Named Stratioticos*. London: Henrie Bynneman.
———. 1590. *An Arithmetical Warlike Treatise Named Stratioticos*. London: Richard Field.
———. 1591. *A Geometrical Practical Treatize Named Pantometria*. London: Abell Ieffes.
Emmius, Ubbo. 1621. *Guilhelmus Ludovicus comes Nassovius*. Groningen: Johannes Sassius.
Errard, Jean. 1584. *Le Premier Livre des Instruments Mathématiques et Mécaniques*. Nancy: Jan-Janson.
———. 1594. *La geometrie et practique generalle d'icelle*. Paris: D. Le Clerc.
Euclid. 1570. *The Elements of Geometrie of the Most Ancient Philosopher Euclide of Megara . . . With a very fruitfull praeface made by M. I. Dee*. Trans. Henry Billingsley. London: Iohn Daye.
Fioravanti, Leonardo. 1678. *Dello specchio di scientia universal*. Venetia: Lattoni.

Galilei, Galileo. 1606. *Le operazioni del compasso geometrico et militare*. Padua.
———. 1890–1909. *Le Opere di Galileo Galilei*. Vol. 20 vol. Firenze: Barbera.
———. 1960. *Discourse on Bodies in Water, Theorem IV, Corollary II, tr. Thomas Salusbury*. London, 1663. Rpt. in facsimile, ed. Stillman Drake. Urbana: University of Illinois Press.
———. 1974. *Discourses on Two New Sciences*. Trans. Stillman Drake. Madison: University of Wisconsin Press.
Gemma, Frisius. 1533. *Libellus de Locorum describendorum ratione*. Antwerp: Arnoldum Birckman.
Hakluyt, Richard. 1582. *Divers Voyages Touching the Discoverie of America, and the Ilands Adiacent Unto the Same, Made First of All by Our Englishmen*. London.
———. 1589. *Principal Navigations, Voyages, and Discoveries of the English Nation*. London.
———. 1598–1600. *The Principal Navigations, Voiages, Traffiques and Discoveries of the English Nation*. 3 Vols. London.
Harvey, Gabriel. 1964. *The Complete Works*, ed. Alexander Grosart, 3 vols. London, 1884; rpt. New York.
Heron of Alexandria. 1589. *Gli artifitiosi et curiosi moti spiritali di Herrone*. Trans. Giambattista Aleotti. Ferrara: V. Baldini.
———. 1592. *Spiritali di Herone Alessandrino*. Trans. Alessandro Giorgi. Urbino: Bartholomeo, e Simone Ragusii.
———. 1616. *Heronis Ctesibii Belopoeeca, hoc est Telifactiva*, ed. Bernardino Baldi. Augustae Vindelicorum [Augsburg]: Davidis Franci.
———. 1962. *Di Herone Alessandrino degli Automati, overo Machine se moventi Libri*. Ed. and Trans. Bernardino Baldi. Venice: Girolamo Porro, 1589; rpt. Milan.
———. 1971. *The Pneumatics of Hero of Alexandria*. Trans. and ed. by Bennet Woodcroft. London: C. Whittingham, 1851; rpt. London: Macdonald & Co.
Hood, Thomas. 1588. *A Copie of the speache: made by the Mathematicall Lecturer unto the Worshipfull Companye present in Gracious Street the 4 of November 1588*. London: Edward Allde.
———. 1590. *The Use of the Celestial Globe in Plano, set foorth in two Hemispheres*. London: Thobie Cooke.
———. 1592. *The Use of Both the Globes, Celestiall, and Terrestriall, most plainely delivered in forme of a Dialogue*. London: Thomas Dawson.
———. 1596. *The Use of the Two Mathematicall Instruments, the crosse Staff. And the Iacobs Staff (1590)*. London: Richard Field for Robert Dexter.
———. 1598. *The Making and Use of the Geometricall Instrument, called a Sector*. London: Iohn Windet.
Hooke, Robert. 1935. *The Diary of Robert Hooke*. Ed. by Henry W. Robinson and Walter Adams. London: Taylor and Francis.
Hopton, Arthur. 1611. *Speculum Topographicum*. London: N[icholas] O[kes] for Simon Waterson.
Lanteri, Giacomo. 1557. *Due dialoghi di M. Iacomo de' Lanteri . . . : ne i quali s'introduce Messer Girolamo Cantanio . . . & Messer Francesco Treuisi . . . 'a ragionare Del modo di disegnare le piante delle fortezze secondo Euclide; et Del modo di comporre i modelli & torre in disegno le piante delle citt'a*. Venice: Vincenzo Valgrisi & Baldessar Contantini.
Lucar, Cyprian. 1588. *Three Bookes of Colloquies Concerning the Arte of Shooting in Great and Small Peeces of Artillerie*. London: [Thomas Dawson] for Iohn Harrison.
———. 1590. *A Treatise Named Lucarsolace*. London: Richard Field for Iohn Harrison.
Marolois, Samuel. 1613–14. *Opera mathematica, ou Oeuvres mathématiques traictans de géométrie, perspective, architecture et fortification*. Hagæ-Comitis: Henrici Hondii.
Metius, A. 1614. *Institutiones Astronomicae et Geographicae*. Amsterdam: Franeker.
———. 1626. *Arithmeticae libri duo et Geometriae*. Leiden: Franeker.
Napier, John. 1616. *A Description of the Admirable Table of Logarithmes*. London: Nicholas Okes.

Norton, Robert. 1624. *Of the Art of Great Artillery*. London: Edw. Allde for Iohn Tap.
———. 1628. *The Gunner Shewing the Whole Practise of Artillery*. London: Humphrey Robinson.
Norwood, Richard. 1639. *Fortification or Architecture Military*. London: Tho. Cotes, for Andrew Crooke.
Parker, Edmund. (*c*.1590–1602). *Notebook*. London: Lambeth Palace Library, MS 280.
Piccolomini, Alessandro. 1547. *In Mechanicas Quaestiones Aristotelis Paraphrasis*. Rome: A. Bladum.
Pigafetta, Filipp. 1581. *Le mechaniche: nelle quali si contiene la vera dottrina di tutti gli istrumenti principali da mover pesi grandissimi con picciola forza*. Venice: Francesco di Franceschi.
Plutarch. 1917. *Plutarch's Lives: Agesilaus and Pompey, Pelopidas and Marcellus*. Trans. Bernadotte Perrin. Cambridge: Harvard University Press.
Polter, Richard. 1605. *The Pathway to Perfect Sayling*. London: Edward Allde for Iohn Tappe.
Ramelli, Agostino. 1588. *Le Diverse et Artificiose Machine*. Paris: casa del Autore.
Ramus, Petrus. 1567. *Prœmium Mathematicum*. Paris: Andream Wechelum.
———. 1590. *The Elementes of Geometrie*. Trans. Thomas Hood. London: Iohn Windet.
Rathborne, Aaron. 1616. *The Surveyor: in foure books*. London: W. Stansby for W. Burre.
Recorde, Robert. 1542. *The Grounde of Artes, Teachyng the Worke and Practice of Arithmetike*. London: Wolf.
———. 1561. *The Pathway to Knowledge, Containing the First Principles of Geometrie*. London: Reginalde Wolfe.
———. 1574. *The Pathewaie to Knowledge Containyng the First Principles of Geometrie*. London: [J. Kingston for] Ihon Harrison.
Ruscelli, Girolamo. 1568. *Precetti della militia moderna . . . tutta l'arte del Bombardiero*. Venice: Marchiò Sessa.
Santbech, Daniel. 1561. *Problematum astronomicorum et geometricorum sectiones septem*. Basel: Henrichum Petri and Petrum Pernam.
Smith, Thomas. 1600. *The Arte of gunnerie: Wherein is set foorth a number of seruiceable secrets, and practical conclusions, belonging to the art of gunnerie, by Arithmeticke Skill to be Accomplished*. London: [Richard Field] for William Ponsonby.
Sprat, Thomas. 1667. *The History of the Royal Society of London*. London: J.R., for J. Martyn.
Stafford, Thomas. 1633. *Pacata Hibernia, Ireland Appeased and Reduced*. London: A[ugustine] M[athewes].
Stevin, Simon. 1586. *De Beghinselen des Waterwichts*. Leiden: Christoffel Plantijn, by Françoyn van Raphelinghen.
———. 1599. *The Haven Finding Art*. Trans. Edward Wright. London: G.B.R.N. and R.B.
———. 1955–66. *The Principal Works of Simon Stevin*. Vol. 5 vols. Amsterdam: C.V. Swets & Zeitlinger.
Stow, John. 1598. *A Survey of London*. London: [John Windet for] Iohn Wolfe.
Taccola, Mariano. 1984. *De Rebus Militaribus (De machinis, 1449)*, ed. Eberhard Knobloch. Saecula Spiritalia Ingenieria e Arte Militare 11. Baden-Baden: V. Koerner.
Tapp, John. 1613. *The Path-Way to Knowledge*. London: T: Purfoot, for Tho: Pauier.
Tartaglia, Niccolò. 1537. *Nova Scientia*. Venice: S. da Sabio.
———. 1546. *Quesiti e Inventioni diverse*. Venice: Lulio.
Ufano, Diego. 1612. *Tratado de la Artilleria*. Brussels: Juan Momarte.
Valla, Giorgio. 1501. *De expetendis de fugiendis rebus opus*. Venice: Aldus.
Worsop, Edward. 1582. *A Discoverie of sundrie errours and faults*. London: Henrie Middleton for Gregorie Seton.
Wright, Edward. 1599a. *Certaine Errors in Navigation*. London: Valentine Sims [and W. White].
———. 1599b. *The Voyage of the Right Honorable the Earle of Cumberland to the Islands Azores A.D. 1589*. London: Valentine Sims.
———. 1610. *Certaine Errors in Navigation*. 2nd ed. London: Felix Kingsto[n].
Wright, Richard. c.1560. *Notes on Gunnery*. London: The Society of Antiquaries of London, MS 94.
Zonca, Vittorio. 1607. *Novo Teatro di Machine et Edificii*. Padua: Pietro Bertelli.

Secondary Sources

Aalbers, Johan, eds. 1987. *De Bloem der Natie. Adel en patriciaat in de Noordelijke Nederlanden*. Meppel: Bloom.

Ackerman, Silke. Cole, Humfrey (d. 1591). In *DNB*. Online at http://www.oxforddnb.com/view/article/5853.

———. ed. 1998. *Humphrey Cole: Mint, Measurement, and Maps in Elizabethan England*. British Museum Occasional Paper 126. London.

Adamson, Ian R. 1976. *Foundation and early History of Gresham College, London, 1596–1704*. Ph.D. dissertation, Cambridge University.

———. 1980. The Administration of Gresham College and its Fluctuating Fortunes as a Scientific Institution in the Seventeenth Century. *History of Education* 9: 13–25.

Alder, Ken. 1997. *Engineering the Revolution: Arms and Enlightenment in France, 1763–181*. Princeton: Princeton University Press.

Alexander, Amir R. 1995. The Imperialist Space of Elizabethan Mathematics. *Studies in the History and Philosophy of Science* 26: 559–591.

———. 2002. *Geometrical Landscapes: The Voyages of Discovery and the Transformation of Mathematical Practice*. Stanford: Stanford University Press.

Anderson, Christy. 2003. The Secrets of Vision in Renaissance England. pp. 323–347 in Massey, *The Treatise on Perspective*.

Applebaum, Wilbur, eds. 2000. *Encyclopedia of the Scientific Revolution*. New York: Garland Publishing.

Apt, A.J. Wright, Edward (bap. 1561, d. 1615). In *DNB*. Online at http://www.oxforddnb.com/view/article/30029.

Archer, Ian. 1991. *The Pursuit of Stability: Social Relations in Elizabethan England*. Cambridge: Cambridge University Press.

Ash, Eric H. 2000. A Perfect and an Absolute Work: Expertise, Authority, and the Rebuilding of Dover Harbor, 1579–1583. *Technology & Culture* 41(2): 239–268.

———. 2004. *Power, Knowledge, and Expertise in Elizabethan England*. Baltimore/London: Johns Hopkins Press.

———. 2010. *Expertise: Practical Knowledge and the Early Modern State*. *Osiris* 25 Chicago: University of Chicago Press.

Baigent, Elizabeth. Ryther, Augustine (*d.* 1593). In *DNB*. Online at http://www.oxforddnb.com/view/article/24428.

Baird, Davis. 2004. *Thing Knowledge: A Philosophy of Scientific Instruments*. Berkeley: University of California Press.

Barnes, Barry. 1982. *T.S. Kuhn and Social Science*. New York: Columbia University Press.

Bawlf, Samuel. 2003. *The Secret Voyage of Sir Francis Drake 1577–1580*. New York: Walker.

Ben-David, Joseph. 1971. *The Scientist's Role in Society: A Comparative Study*. Englewood Cliffs: Prentice Hall.

Bennett, James A. 1986. The Mechanic's Philosophy and the Mechanical Philosophy. *History of Science* 24: 1–28.

———. 1991. The Challenge of practical mathematics. In *Science, Belief, and Popular Culture in Renaissance Europe*, ed. Stephen Pumfrey et al., 176–190. Manchester: Manchester University Press.

———. 1998. Practical Geometry and Operative Knowledge. *Configurations* 6: 195–222.

———. 2003. Knowing and Doing in the Sixteenth Century: What Were Instruments for? *British Journal for the History of Science* 36: 129–150.

———. 2007. Wind-Gun, Air-Gun or Pop-Gun: The Fortunes of a Philosophical Instrument. In *The Mindful Hand*, ed. Roberts et al., 221–246. Amsterdam: Koninkliijke Nederlandse Akademie van Wetenschappen.

———. 2011. Early Modern Mathematical Instruments. *Isis* 102(4): 697–705.

Bergsma, W. 1998. Willem Lodewijk en het Leeuwarder hofleven. *It Beaken* 60: 199–201, 215–227.

Bernal, J.D. 1954. *Science in History*. London: Watts & Co..

Biagioli, Mario. 1989. The Social Status of Italian Mathematicians, 1450–1600. *History of Science* 27: 41–95.

———. 1993. *Galileo, Courtier: The Practice of Science in the Culture of Absolutism*. Chicago: University of Chicago Press.

———. 2007. *Galileo's Instruments of Credit: Telescopes, Images, Secrecy*. Chicago: University of Chicago Press.

Bloor, David. 1976. *Knowledge and Social Imagery*. London: Routledge/K. Paul.

Boas, Marie. 1949. Hero's Pneumatica. A Study of its Transmission and Influence. *Isis* 40: 38–48, repr. in M.B. Hall, *The Mechanical Philosophy*. 1981.

Borkenau, Franz. 1934. *Der Ubergang vom feudalen zum burgerlichen Weltbild. Studien zur Geschichte der Manufakturperiode*. Paris: F. Alcan.

Bound, Mensun, eds. 1998. *Excavating Ships of War*. Oswestry: A. Nelson.

Brewer, J.S., and William, Bullen. eds. 1601–1603. *Calendar of the Carew Manuscripts preserved in the archiepiscopal library at Lambeth*. Vol. 4. London: Green, Reader & Dyer, 1870.

Brotton, Jerry. 1997. *Trading Territories. Mapping the Early Modern World*. London: Reaktion Books.

Brown, Joyce. 1979. *Mathematical Instrument-Makers in the Grocers Company 1688–1800*. London: Science Museum.

Brown, Ruth R. 1990. Comment on The Tal-y-Bont Gunner's Rule. *Journal of the Ordnance Society* 2: 71–72.

Bruff, Lawrence L. 1892. *Exterior Ballistics, Gun Construction, and U.S. Seacoast Guns*. West Point: United States Military Academy Press.

Bukharin, Nikolai. 1931. *Science at the Cross Roads; Papers Presented to the International Congress of the History of Science and Technology Held in London from June 20th to July 3rd, 1931 by the Delegates of the U.S.S.R.* London: Kniga Ltd..

Burke, Peter. 1978. *Popular Culture in Early Modern Europe*. London: Temple Smith.

Burtt, Edwin A. 1924. *The Metaphysical Foundations of Modern Science*. London: K. Paul, Trench, Trübner.

Bush, Vannevar. 1945. *Science, the Endless Frontier*. Washington, DC: G.P.O.

Butterfield, Herbert. 1949. *The Origins of Modern Science, 1300–1800*. London: Bell.

Büttner, Jochen, Peter, Damerow, Jürgen, Renn, and Matthias, Schemmel. 2003. The Challenging Images of Artillery: Practical knowledge at the roots of the scientific revolution. In *The Power of Images*, Lefèvre, 3–27.

Carteron, Henri. 1979. *La Notion de force dans la système d'Aristote*. New York: Garland.

Charleston, R.J. 1985. *English Glass and the Glass Used in England, Circa 400–1940*. London: George Allen & Unwin.

Chartier, Roger. 1988. *Cultural History: Between Practices and Representation*. Ithaca: Cornhell University Press.

Clagett, Marshall, eds. 1959. *Critical Problems in the History of Science*. Madison: University of Wisconsin Press.

Clifton, Gloria, and Gerard L.E. Turner, eds. 1995. *Directory of British Scientific Instrument Makers 1550–1851*. London: Philip Wilson Publishers.

Clucas, Stephen, ed. 2006. *John Dee: Interdisciplinary Studies in English Renaissance Thought*. Dordrecht: Springer.

Clulee, Nicholas H. 1977. Astrology, Magic, and Optics: Facets of John Dee's Early Natural Philosophy. *Renaissance Quarterly* 30: 632–680.

———. 1988. *John Dee's Natural Philosophy: Between Science and Religion*. London/New York: Routledge.

Cockle, Maurice J.D. 1957. *A Bibliography of Military Books Up to 1642*. 2nd ed. London: Holland Press.

Cohen, H. Floris. 1994. *The Scientific Revolution. A Historiographical Inquiry*. Chicago: University of Chicago Press.

———. 2010. *How Modern Science Came into the World: Four Civilizations, One Seventeenth-Century Breakthrough*. Amsterdam: Amsterdam University Press.

Colvin, H.M., John Summerson, Martin Biddle, J.R. Hale, and Marcus Merriman, eds. 1982. *The History of the King's Works, vol. 4, 1485–1660 part II*. London: Her Majesty's STationary Office.

Cordoba, Ricardo, eds. 2013. *Craft Treatises and Handbooks: The Dissemination of Technical Knowledge in the Middle Ages*, De Diversis Artibus 91. Turnhout: Brepols Publishers.

Cormack, Lesley B. 1991a. 'Good Fences Make Good Neighbors': Geography as Self-Definition in Early Modern England. *Isis* 82: 639–661.

———. 1991b. Twisting the Lion's Tail: Practice and Theory at the Court of Henry Prince of Wales. In *Patronage and Institution: Science, Technology, and Medicine at the European Court, 1500–1750*, ed. Bruce T. Moran, 67–84. Woodbridge: Boydell Press.

———. 1997. *Charting an Empire. Geography at the English Universities 1580–1620*. Chicago: University of Chicago Press.

———. 2003. The Grounde of Artes: Robert Recorde and the role of the Muscovy Company in an English mathematical Renaissance. *Proceedings of the Canadian Society for the History and Philosophy of Mathematics* 16: 132–138.

———. 2009. Glob(al) Visions: Globes and their Publics in early modern Europe. In *Making Publics in Early Modern Europe: People, Things, Forms of Knowledge*, ed. Paul Yachnin and BronwenWilson, 138–156. New York: Routledge University Press.

Cowan, Brian. 2005. *The Social Life of Coffee: The Emergence of the English Coffeehouse*. New Haven: Yale University Press.

Crino, Anna Maria, and Helen Wallis. 1987. New Researches on the Molyneux Globes. *Der Globusfreund* 35(37): 11–20.

Croiset van Uchelen, A.R.A., Koert van der Horst, and Günter G. Schilder, eds. 1989. *Theatrum Orbis Librorum*. Utrecht: HES Publishers.

Crosland, Maurice P., eds. 1975. *The Emergence of Science in Western Europe*. London: Science History Publications.

Cunningham, Andrew. 1988. Getting the Game Right: Some Plain Words on the Identity and Invention of Science. *Studies in History and Philosophy of Science Part A* 19(3): 365–389.

———. 1991. How the *Principia* Got its Name; or, Taking Natural Philosophy Seriously. *History of Science* 29: 377–392.

Cunningham, Andrew, and Perry Williams. 1993. De-Centring the 'Big Picture': The Origins of Modern Science and the Modern Origins of Science. *British Journal for the History of Science* 26: 407–432.

de Clercq, P. 1988. Science at Court: The Eighteenth-Century Cabinet of Scientific Instruments and Models of the Dutch Stadholders. *Annals of Science* 45: 113–152.

de La Croix, Horst. 1963. The Literature on Fortification in Renaissance Italy. *Technology & Culture* 4: 30–50.

Dear, Peter. 1991. The Church and the New Philosophy. In *Science, Culture, and Popular Belief in Renaissance Europe*, ed. Stephen Pumfrey et al., 119–139. Manchester: Manchester University Press.

———. 1995. *Discipline and Experience: The Mathematical Way in the Scientific Revolution*. Chicago: University of Chicago Press.

———. 2001a. Religion, Science and Natural Philosophy: Thoughts on Cunningham's Thesis. *Studies in History and Philosophy of Science Part A* 32(2): 377–386.

———. 2001b. *Revolutionizing the Sciences: European Knowledge and its Ambitions, 1500–1700*. Princeton: Princeton University Press.

Dijksterhuis, E.J. 1961. *The Mechanization of the World Picture*. Oxford: Oxford University Press.

Dijksterhuis, Fokko Jan. 1999. Vader en Zoon. Over Constantijn en Christiaan Huygens. *Bzzlletin* 28: 18–22.

———. 2003. Stampioen Jr, Jan Janszoon (1610-after1689). In *Dictionary of Seventeenth and Eighteenth-Century Dutch Philosophers*, ed. W. van Bunge et al., 938–940. Bristol: Thoemmes Press.

———. 2004. *Lenses and Waves. Christiaan Huygens and the Mathematical Science of Optics in the Seventeenth Century*, Archimedes. Vol. 9. Dordrecht: Springer.

———. 2007. Constructive Thinking. A Case for Dioptrics. In *The Mindful Hand*, ed. Roberts et al., 59–82. Amsterdam: Koninkliijke Nederlandse Akademie van Wetenschappen.

———. 2009. Fit to Measure. 'Bequamheit' in Mathematics in the Dutch Republic. In *Public Offices, Personal Demands*, ed. Hartman et al., 80–100. Newcastle upon Tyne: Cambridge Scholars Publishing.

———. 2011. Moving Around the Ellipse. Conic Sections in Leiden (1620–1660). In *Silent Messengers*, ed. S. Dupré and C.H. Lüthy, 89–124. Berlin: Transaction Publishers.

Dijkstra, A.F.B. 2007. *Het vinden van Oost en West*. M.A. thesis, University of Groningen.

Dijkstra, A. 2011. A Wonderful Little Book. The Dissertatio Astronomica by Johannes Phocylides Holwarda (1618–1651). In *Centres and Cycles*, ed. L. Roberts, 73–100. London: LIT Verlag.

Dijkstra, Arjen. 2012. *Between Academics and Idiots: A Cultural History of Mathematics in the Dutch Province of Friesland (1600–1700)*. Ph.D. thesis, Universiteit Twente.

Dobres, Marcia-Anne. 1999. Technology's Links and Chaînes: The Processural Unfolding of Technique and Technician. In *Social Dynamics of Technology*, ed. C.R. Hoffman and M.-A. Dobres, 124–146. Washington, DC: Smithsonian Institution Press.

Donagan, Barbara. 1995. Halcyon Days and the Literature of War: England's Military Education before 1642. *Past & Present* 147: 65–100.

Douglas, Mary. 1970. *Natural Symbols: Explorations in Cosmology*. London: Barrie and Rockliff.

Drachmann, A.G. Hero of Alexandria. *DSB* 6: 310–314.

———. 1963. *The Mechanical Technology of Greek and Roman Antiquity*. Copenhagen: Hafner Publishing Company.

Drake, Stillman, and I.E. Drabkin. 1969. *Mechanics in Sixteenth Century Italy*. Madison: University of Wisconsin Press.

———. 1970. Early Science and the Printed Book: The Spread of Science Beyond the Universities. *Renaissance and Reformation* 6: 43–52.

———. 1978. *Galileo at Work: His Scientific Biography*. Chicago: University of Chicago Press.

Duffy, Christopher. 1979. *Siege Warfare; the Fortress in the Early Modern World*. London: Routledge/Kegan Paul.

Duhem, Pierre. 1905–6. *Les Origines de la Statique*. Vol. 2 vols. Paris: A. Hermann.

Dupré, Sven. 2003. The Dioptrics of Refractive Dials in the Sixteenth Century. *Nuncius* 18: 53–57.

———. 2005. Ausonio's Mirrors and Galileo's Lenses: The Telescope and Sixteenth-Century Practical Optical Knowledge. *Galilaeana: Journal of Galilean Studies* 2: 145–180.

———. 2006. Visualization in Renaissance optics: the function of geometrical diagrams and pictures in the transmission of practical knowledge. In *Transmitting Knowledge*, ed. S. Kusukawa and I. Maclean, 11–39. Oxford: Oxford University Press.

Dupré, Sven, and Christoph Lüthy, eds. 2011. *Silent Messengers. The Circulation of Material Objects of Knowledge in the Early Modern Low Countries*. Berlin: Lit.

Eagleton, Catherine. 2006. Medieval Sundials and Manuscript Sources: The Transmission of Information About the Navicula and the Organum Ptolemei in Fifteenth-Century Europe. In *Transmitting Knowledge*, ed. S. Kusukawa and I. Maclean, 41–71. Oxford: Oxford University Press.

Eamon, William. 1994. *Science and the Secrets of Nature: Books of Secrets in Medieval and Early Modern Culture*. Princeton: Princeton University Press.

Ede, Andrew, and Lesley Cormack, eds. 2007. *A History of Science in Society: A Reader*. Peterborough: Westview Press.

Etzkowitz, Henry, Andrew Webster, and Peter Healey, eds. 1998. *Capitalizing Knowledge: New Intersections of Industry and Academia*. New York: State University of New York Press.

Falls, Cyril B. 1950. *Elizabeth's Irish Wars*. London: Methuen.
Favaro, Antonio. 1886. La libreria di Galileo Galilei. *Bullettino di Bibliografia e di Storia delle Scienze Mathematiche e Fisiche*. 19: 219–290.
Feingold, Mordechai. 1984. *The Mathematician's Apprenticeship: Science, Universities and Society in England, 1560–1640*. Cambridge: Cambridge University Press.
Ferguson Eugene S., ed. and Martha T. Gnudi. 1976. Trans. *The Various and Ingenious Machines of Agostino Ramelli (1588)*. Baltimore: John Hopkins University Press.
Feyerabend, Paul K. 1993. *Against Method*. 3rd ed. London: New Left Books.
Findlen, Paula. 1994. *Possessing Nature: Museums, Collecting, and Scientific Culture in Early Modern Italy*. Berkeley: University of California Press.
Fiocca, Alessandra, eds. 1998a. *Giambattista Aleotti e gli ingegneri del Rinascimento*. Florence: L.S. Olschki.
———. 1998b. Giambattista Aleotti e la 'Scienza et arte delle acque. In *Giambattista Aleotti*, ed. A. Fiocca, 47–101. Firenze: L.S. Olschki.
Fiore, F.P. 1978. *Cittá e Macchine del '400 nei Disegni di Francesco di Giorgio Martini*, Studi of the Accademia toscana di scienze e lettere La Colombaria. Vol. 49. Florence: L.S. Olschki.
Fisher, R.M. 1974. William Crashawe and the Middle Temple Globes 1605–15. *Geographical Journal* 140: 105–112.
Foulché-Delbosch, R. 1898. Mechanica de Aristoteles. *Revue Hispanique* 5: 365–405.
France-Lanord, Albert, eds. 1979. *Le Premier livre des instruments mathématiques méchaniques*. Paris: Berger-Levrault.
Gabriel, Astrik L. 1968. *A Summary Catalogue of Microfilms of One Thousand Scientific Manuscripts in the Ambrosiana Library, Milan*. Notre Dame: The Mediaeval Institute.
Gadd, Ian Anders, eds. 2006. *Guilds and Association in Europe, 900–1900*. London: Centre for Metropolitan History.
Gadd, Ian Anders, and Patrick Wallis, eds. 2002. *Guilds, Society & Economy in London, 1450–1800*. London: Centre for Metropolitan History.
Gamba, Enrico, and Vico Montebelli. 1988. *Le scienze a Urbino nel Tardo Rinascimento*. Urbino: QuattroVenti.
Garber, Daniel, eds. 2002. *La Rivoluzione Scientifica*. Rome: Instituto della Enciclopedia Italiana.
Gasser, Les, and N. Huhns, eds. 1989. *Distributed Artificial Intelligence*. London: Pitman.
Gatto, Romano. 2015. La *Meccanica* di Erone nel Rinascimento. In *Scienze e rappresentazioni: Saggi in onore di Pierre Souffrin*, ed. Pierre Caye, Romano Nanni, and Pier Daniele Napolitani. Florence: Olschki.
Gaukroger, Stephen, eds. 1980. *Descartes: Philosophy, Mathematics and Physics*. Brighton/Sussex: Harvester Press.
———. 1995. *Descartes: An Intellectual Biography*. Oxford: Clarendon Press.
———. 2000. The Foundational Role of Hydrostatics and Statics in Descartes' Natural Philosophy. In *Descartes' Natural Philosophy*, ed. S. Gaukroger, J.A. Schuster, and J. Sutton, 60–80. London: Routledge.
Gaukroger, Stephen, and John A. Schuster. 2002. The hydrostatic paradox and the origins of Cartesian dynamic. *Studies in History and Philosophy of Science Part A* 33(3): 535–572.
Gaukroger, Stephen, John Schuster, and John Sutton, eds. 2000. *Descartes' Natural Philosophy*. London: Routledge.
Gauvin, Jean-François. 2006. Artisans, Machines and Descartes's *Organon*. *History of Science* 44(2): 187–216.
Gerbino, Anthony, and Stephen Johnston. 2009. *Compass and Rule: Architecture as Mathematical Practice in England 1500–1750*. New Haven: Yale University Press.
Giacomoni, Paola, eds. 2011. *Jacopo Aconcio: Trattato sulle fortificazioni*, Istituto Nazionale di Studi sul Rinascimento, Studi e Testi. Vol. 48. Firenze: Leo S. Olschki.
Glover, Winifred. 1998. The Spanish Armada Wrecks of Ireland. In *Excavating Ships of War*, ed. M. Bound, 51–63. Oswestry: A. Nelson.
Godfrey, Eleanor S. 1975. *The Development of English Glassmaking 1560–1640*. Chapel Hill: University of North Carolina Press.

Goffman, Erving. 1959. *The Presentation of Self in Everyday Life*. Garden City: Doubleday.

———. 1979. *Frame Analysis: An Essay on the Organization of Experience*. New York: Harper & Row.

Golinski, Jan. 1998. *Making Natural Knowledge: Constructivism and the History of Science*. Cambridge: Cambridge University Press.

Gorham, Geoffrey, and Benjamin Hill, eds. 2015. *The Language of Nature: Reassessing the Mathematization of Natural Philosophy*. Minneapolis: University of Minnesota Press.

Graham, Loren. 1985. Socio-political Roots of Boris Hessen: Soviet Marxism and the History of Science. *Social Studies of Science* 15: 705–722.

———. 1993. *Science in Russia and the Soviet Union: A Short History*. Cambridge: Cambridge University Press.

Grant, Edward. 1971. Henricus Aristippus, William of Moerbeke and Two Alleged Medieval Translations of Hero's Pneumatica. *Speculum* 96: 659–660.

Graves, Michael A.R. 1998. *William Cecil, Lord Burghley*. London/New York: Longman.

Green, Jeremy N. 1990. Further Information on Gunner's Rules or Tally Sticks. *Journal of the Ordnance Society* 2: 25–32.

Grendler, Marcella. 1980. A Greek Collection in Padua: The Library of Gian Vincenzo Pinelli (1535–1601). *Renaissance Quarterly* 33: 402–405.

Guilmartin, John F. Jr. 1989. Ballistics in the Black Powder Era. In *British Naval Armaments*, ed. R.D. Smith, 73–98. London: Trustees of the Royal Armouries.

Hadden, Richard. 1994. *On the Shoulders of Merchants: Exchange and the Mathematical Conception of Nature in Early Modern Europe*. Albany: State University of New York Press.

Hale, J.R. 1977. *Renaissance Fortification: Art or Engineering?* London: Thames and Hudson.

Hall, A. 1952. *Rupert. Ballistics in the Seventeenth Century: a Study in the Relations of Science and War with Reference Principally to England*. Cambridge: Cambridge University Press.

———. 1959. The Scholar and the Craftsman in the Scientific Revolution. In *Critical Problems in the History of Science*, ed. M. Clagett, 3–23. Madison: University of Wisconsin Press.

Hall, Marie Boas. 1962. *The Scientific Renaissance 1450–1630*. New York: Harper & Brothers.

———. 1971. A Note on the Text and Its Accompanying Illustrations. In Hero of Alexandria, *The Pneumatics*. Bennet Woodcroft, ed., Joseph George Greenwood., Trans. London: Macdonald & Company.

———. 1981. *The Mechanical Philosophy*. New York: Arno Press.

Halliwell-Phillipps, J.O. 1839. *Rara Mathematica; Or, A Collection of Treatises on the Mathematics and Subjects Connected with Them*. London: J.W. Parker.

Hantsche, Irmgard, eds. 1996. *Der 'mathematicus': Zur Entwicklung und Bedeutung einer neun Berufsgruppe in der Zeit Gerhard Mercators*. Bochum: Brockmeyer.

Harkness, Deborah E. 1999. *John Dee's Conversations with Angels: Cabala, Alchemy and the End of Nature*. Cambridge: Cambridge University Press.

———. 2007. *The Jewel House. Elizabethan London and the Scientific Revolution*. New Haven: Yale University Press.

Harrison, Peter. 2000. The Influence of Cartesian Cosmology in England. In *Descartes' Natural Philosophy*, ed. S. Gaukroger, J.A. Schuster, and J. Sutton, 168–192. London: Routledge.

———. 2002. Voluntarism and Early Modern science. *History of Science* 40: 63–89.

———. 2005. Physico-Theology and the Mixed Sciences: The Role of Theology in Early Modern Natural Philosophy. In *The Science of Nature*, ed. J.A. Schuster and P.R. Anstey, 165–183. Dordrecht: Springer.

Hartman, Jan, Jaap Nieuwstraten, and Michel Reinders, eds. 2009. *Public Offices, Personal Demands. Capability in Governance in the Seventeenth-Century Dutch Republic*. Newcastle: Cambridge Scholars Publishing.

Haskins, Charles Homer. 1960. *Studies in the History of Medieval Science*. Cambridge, MA: Harvard University Press, 1924; rpt.

Hattab, Helen. 2005. From Mechanics to Mechanism: The *Quaestiones Mechanicae* and Descartes' Physics. In *The Science of Nature*, ed. J.A. Schuster and P.R. Anstey, 99–129. Dordrecht: Springer.

Heal, Felicity and Clive Holmes. 2002. The Economic Patronage of William Cecil. In *Patronage, Culture and Power: The Early Cecils*, ed. Pauline Croft, 199–299. New Haven: Yale University Press.

Helgerson, Richard. 1992. *Forms of Nationhood. The Elizabethan Writing of England*. Chicago: University of Chicago Press.

Heller, Henry. 2003. *Anti-Italianism in Sixteenth Century France*. Toronto: University of Toronto Press.

Henry, John. 2001. *The Scientific Revolution and the Origins of Modern Science*. 3rd ed. Houndmills/Basingstoke: Palgrave.

Hessen, Boris. 1931. The Social and Economic Roots of Newton's Principia. In *Science at the Cross Roads*, ed. N. Bukharin, 149–213. London: Kniga.

Higton, Hester K. Hood, Thomas (*bap.* 1556, *d.* 1620). In *DSB*.

———. Gunter, Edmund. 1581–1626. In *DNB*. Online at http://www.oxforddnb.com/view/article/11751.

———. Allen, Elias. (*c.*1588–1653). In *DNB*. Online at http://www.oxforddnb.com/view/article/37108.

———. 2001. Does Using an Instrument Make You Mathematical? Mathematical Practitioners of the 17th Century. *Endeavour* 25(1): 18–22.

———. 2004. Portrait of an instrument-Maker: Wenseslaus Hollar's Engraving of Elias Allen. *British Journal for the History of Science* 27(2): 146–166.

Hildred, Alexzandra. 1997. The Material Culture of the Mary Rose (1545) as Fighting Vessel: The Uses of Wood. In *Artefacts from Wrecks*, ed. M. Redknap, 51–72. Oxford: Oxbow Books.

———. 2011. *Weapons of Warre: The Armaments of the Mary Rose*. Portsmouth: Mary Rose Trust.

Hill, Christopher. 1965. *Intellectual Origins of the English Revolution*. Oxford: Clarendon Press.

Hill, Katherine. 1998. 'Juglers or Schollers?': Negotiating the Role of a Mathematical Practitioner. *British Journal for the History of Science* 31: 253–274.

Hoffman, Christopher R., and Marcia-Anne Dobres. 1999a. Conclusion: Making Material Culture, Making Culture Material. In *Social Dynamics of Technology*, ed. C.R. Hoffman and M.-A. Dobres, 209–222. Washington , DC: Smithsonian Institution Press.

———, eds. 1999b. *The Social Dynamics of Technology: Practice, Politics, and World Views*. Washington, DC: Smithsonian Institution Press.

Hofmann, Joseph Ehrenfried. 1962. *Frans van Schooten der Jüngere*. Wiesbaden: Steiner.

Hooykaas, R. 1987. The Rise of Modern Science: When and Why? *British Journal for the History of Science* 20: 463–467.

Hotson, H. 2007. *Commonplace Learning. Ramism and Its German Ramifications, 1543–1630*. Oxford: Oxford University Press.

Huff, Toby. 2010. *Intellectual Curiosity and the Scientific Revolution. A Global Perspective*. Cambridge: Cambridge University Press.

Ingalls, James M. 1885. *Exterior Ballistics*. Fort Monroe: Artillery School.

———. 1886. *Exterior Ballistics in Plane of Fire*. New York: D. Van Nostrand.

Israel, J. 1998. *The Dutch Republic. Its Rise, Greatness, and Fall 1477–1806*. Oxford: Oxford University Press.

James, Kathryn. 2011. Reading Numbers in Early Modern England. *British Society for the History of Mathematics [BSHM] Bulletin* 26(1): 1–16.

Jardine, Lisa. 2004. *The Curious Life of Robert Hooke. The Man Who Measured London*. New York: HarperCollins.

Johnson, Francis R. 1940. Gresham College: Precursor of the Royal Society. *Journal of the History of Ideas* 1(4): 413–438.

Johnston, Stephen. Recorde, Robert. (*c.*1512–1558). In *DNB*. Online at http://www.oxforddnb.com/view/article/23241.

———. Digges, Thomas (*c.*1546–1595). In *DNB*. Online at http://www.oxforddnb.com/view/article/7639.

———. 1991. Mathematical Practitioners and Instruments in Elizabethan England. *Annals of Science* 48(4): 319–344.

———. 1994. *Making Mathematical Practice: Gentlemen, Practitioners and Artisans in Elizabethan England*. Ph.D. dissertation, Cambridge University.
———. 1996. The Identity of the Mathematical Practitioner in 16th Century England. In *Der 'mathematicus'*, ed. I. Hantsche, 93–120. Bochum: Brockmeyer.
———. 2004. Theory, Theoric, Practice: Mathematics and Magnetism in Elizabethan England. *Journal de la Renaissance* 2: 53–62.
———. 2006. Like Father, Like Son? John Dee, Thomas Digges and the Identity of the Mathematician. In *John Dee: Interdisciplinary Studies in English Renaissance Thought*, ed. Clucas, Stephen, 65–84. Dordrecht: Springer.
———. 2013. A Revised Bibliography of William Bourne. Online at http://www.mhs.ox.ac.uk/staff/saj/bourne/.
Jones, Richard Foster. 1936. *Ancients and Moderns: A Study of the Rise of the Scientific Movement in Seventeenth-Century England*. Berkeley: University of California Press.
Jones, Matthew L. 2006. *The Good Life in the Scientific Revolution. Descartes, Pascal, Leibniz, and the Cultivation of Virtue*. Chicago: University of Chicago Press.
Kaunzner, Wolfgang. Briggs, Henry (*bap.* 1561, *d.* 1631). In *DNB*. Online at http://www.oxforddnb.com/view/article/3407.
Keller, Alex G. Aconcio, Jacopo (*c.*1520–1566/7?). In *DNB*. Online at http://www.oxforddnb.com/view/article/69.
———. 1950. Zilsel, the Artisans, and the Idea of Progress in the Renaissance. *Journal of the History of Ideas* 11(2): 235–240.
———. 1967. Pneumatics, Automata and the Vacuum in the Work of Giambattista Aleotti. *British Journal for the History of Science* 3: 338–347.
———. 1973. The Missing Years of Jacques Besson, Inventor of Machines, Teacher of Mathematics, Distiller of Oils and Huguenot Pastor. *Technology & Culture* 14: 28–39.
———. 1975. Mathematicians Mechanics and Experimental Machines in Northern Italy in the Sixteenth Century. In *The Emergence of Science*, ed. M.P. Crosland, 15–34. New York: Science History Publications.
———. 1978. Renaissance Theaters of Machines. *Technology & Culture* 19: 495–508.
Konstam, A. 1989. A Gunner's Rule from the 'Bronze Bell' Wreck, Tal-y-Bont. *Journal of the Ordnance Society* 1: 23–26.
Kooijmans, L. 1987. Patriciaat en aristocratisering in Holland tijdens de zeventiende en achttiende eeuw. In *De Bloem der Natie*, ed. Aalbers, 98–103.
Korey, Michael. 2007. *The Geometry of Power: Mathematical Instruments and Princely Mechanics Around 1600*. Munich: Deutscher Kunstverlag.
Koyré, Alexandre. 1957. *From a Closed World to an Infinite Universe*. Baltimore: Johns Hopkins University Press.
Kristeller, Paul Oskar. 1965–67. *Iter Italicum*. Vol. 2 vols. London/Leiden: The Warburg Institute.
Kuhn, Thomas. 1957. *The Copernican Revolution*. Cambridge, MA: Harvard University Press.
———. 1962. *The Structure of Scientific Revolutions*. Chicago: University of Chicago Press.
———. 1977. *The Essential Tension: Selected Studies in Scientific Tradition and Change*. Chicago: University of Chicago Press.
Kusukawa, Sachiko, and Ian Maclean, eds. 2006. *Transmitting Knowledge: Words, Images and Instruments in Early Modern Europe*. Oxford: Oxford University Press.
Lacaita, James Philip, eds. 1879. *Catalogue of the Library at Chatsworth*. Vol. 4 vols. London: Chiswick Press.
Laird, W.R. 1986. The Scope of Renaissance Mechanics. *Osiris* 2: 43–68.
Lamberini, Daniela. 1998. Cultura Ingegneristica nel Granducato di Toscana ai tempi dell' Aleotti. In *Giambattista Aleotti*, ed. A. Fiocca, 293–308. Firenze: L.S. Olschki.
Langins, Janis. 2004. *Conserving the Enlightenment: French Military Engineering from Vauban to the Revolution*. Cambridge, MA: MIT Press.
Latour, Bruno. 2004. Why has Critique Run out of Steam? From Matters of Fact to Matters of Concern. *Critical Inquiry* 30: 225–248.

———. 2005. From Realpolitik to Ding Politik, or How to Make Things Public. In *Making Things Public*, ed. B. Latour and P. Weidbel, 14–41. Cambridge, MA: MIT Press.

Latour, Bruno, and Peter Weidbel, eds. 2005. *Making Things Public: Atmospheres of Democracy*. Cambridge, MA: MIT Press.

Layton, Edwin T. 1971. Mirror-Image Twins: The Communities of Science and Technology in 19th-century America. *Technology & Culture* 12: 562–580.

Lefèvre, Wolfgang, Jürgen Renn, and Urs Schoepflin, eds. 2003. *The Power of Images in Early Modern Science*. Basel: Birkhäuser.

LeGrand, Homer. 1990. *E., ed. Experimental Inquiries: Historical, Philosophical and Social Studies of Experiment*. Dordrecht: Reidel.

Levere, Trevor, and G.L'.E. Turner. 2002. *Discussing Chemistry and Steam: The Minutes of a Coffeehouse Philosophical Society 1780–1787*. Oxford: Oxford University Press.

Lindberg, David, and Robert Westman, eds. 1990. *Reappraisals of the Scientific Revolution*. Cambridge: Cambridge University Press.

Livingstone, David. 1990. Geography. In *Companion to the History of Modern Science*, ed. R.C. Olby et al., 743–760. London: Routledge.

Long, Pamela O. 2001. *Openness, Secrecy, Authorship: Technical Arts and the Culture of Knowledge from Antiquity to the Renaissance*. Baltimore: Johns Hopkins University Press.

———. 2011. *Artisan/Practitioners and the Rise of the New Sciences, 1400–1600*. Corvallis: Oregon State University Press.

Lynch, William T. 2006. Surveying and the Cromwellian Reconquest of Ireland. In *Instrumental in War*, ed. Steven Walton, 47–84. Leiden: Brill.

Maffioli, Cesare S. 2013. A Fruitful Exchange/Conflict: Engineers and Mathematicians in Early Modern Italy. *Annals of Science* 70(2): 197–228.

Mahoney, Michael S. 1980. The Beginnings of Algebraic Thought in the Seventeenth Century. In *Descartes*, ed. S. Gaukroger, 141–155. Brighton: Harvester Press.

Maltese, Corrado, eds. 1967. *Francesco di Giorgio Martini, Trattati di Architettura, Ingenieria e Arte Militare, Trattati di architettura 3*. Il Polifilo: Milan.

Marani, Pietro C. 1984. *Disegni di fortificazioni da Leonardo a Michelangelo*. Florence: Cantini.

Marchis, Vittorio, and Luisa Dolza. 1998. L'Acqua, i numeri, le macchine: al sorgere dell' ingegneria idraulica moderna. In *Giambattista Aleotti*, ed. A. Fiocca, 207–221. Firenze: L.S. Olschki.

Marr, Alexander. 2004. Understanding automata in the late Renaissance. *Journal de la Renaissance* 2: 205–222.

Martens, Rhonda. 2000. *Kepler's Philosophy and the New Philosophy*. Princeton: Princeton University Press.

Martin, R. Julian. 1992. *Francis Bacon, The State, and the Reform of Natural Philosophy*. Cambridge: Cambridge University Press.

Martin, Colin. 2001. De-Particularizing the Particular: Approaches to the Investigation of Well-Documented Post-Medieval Shipwrecks. *World Archaeology* 32: 383–399.

Massey, Lyle, eds. 2003. *The Treatise on Perspective: Published and Unpublished*. New Haven: Yale University Press.

Maxwell, Susan. 2004. Emery Molyneux. In *DNB*, Vol. 38, 556–557. Oxford: Oxford University Press.

Mayr, Otto. 1986. *Authority, Liberty and Automatic Machinery in Early Modern Europe*. Baltimore: Johns Hopkins Press.

Mazzolini, Renato G., eds. 1993. *Non-Verbal Communication in Science Prior to 1900*. Florence: Olschki.

McComish, W.A. 1969. The Survival of the Irish Castle in an Age of Cannon. *The Irish Sword* 9: 18.

McConica, James K., ed. 1986. *The Collegiate University, The History of the University of Oxford 3*. Oxford: Oxford University Press.

Meroi, Fabrizio, and Claudio Pogliano, eds. 2001. *Immagini per conoscere: Dal Rinascimento al Rivoluzione Scientifica*. Florence: Leo S. Olschki.

Merton, Robert K. 1938. Science, Technology and Society in Seventeenth-Century England. *Osiris* 4(2): 360–632.

———. 1970. *Science, Technology and Society in Seventeenth-Century England. 2^d ed.* New York: Harper & Row.

Molhuysen, P. 1913. *Bronnen tot de geschiedenis der Leidsche Universiteit*, Rijksgeschiedkundige Publicatiën 20. Vol. vol. 1. The Hague: Nijhoff.

Molland, A.G. 1974. Roger Bacon as Magician. *Traditio* 30: 445–460.

Moran, Bruce T., eds. 1991. *Patronage and Institutions: Science, technology, and medicine at the European court, 1500–1750*. Rochester: Boydell Press.

Mörke, O. 2007. *Wilhelm von Oranien (1533–1584). Fürst und "Vater" der Republik*, Urban Taschenbücher 609. Stuttgart: Kohlhammer.

Morrison-Low, Alison D., Sven Dupré, Stephen Johnston, and Giorgio Strano, eds. 2012. *From Earth-Bound to Satellite. Telescopes, Skills and Networks*, History of Science and Medicine Library–Scientific Instruments and Collections 23/2. Leiden: Brill.

Moscheo, Rosario. 1988. *Francesco Maurolico tra Rinascimento e scienza galileiana*. Messina: Società Messinese di Storia Patria.

Mosley, Adam. 2006. Objects of knowledge: Mathematics and models in sixteenth-century cosmology and astronomy. In *Transmitting Knowledge*, ed. S. Kusukawa and I. Maclean, 41–71. Oxford: Oxford University Press.

Mulder, H.M. 1990. Pure, Mixed and Applied Mathematics: The Changing Perception of Mathematics Through History. *Nieuw Archief voor Wiskunde* 8: 27–41.

Napolitani, Pier Daniele. 2002. Mechanicae artes. In *Francisci Maurolyci Opera Mathematica*, eds. Veronica Gavagna and Pier Daniele Napolitani. Pisa: Dipartimento di Mathematica dell'Università di Pisa. Online at http://www.maurolico.it/Maurolico/sezione.html?path=12.

Neal, Katherine. 1999. The Rhetoric of Utility: Avoiding Occult Associations for Mathematics through Profitability and Pleasure. *History of Science* 37(2): 151–178.

———. 2002. *[Also see Hill, Katherine] From discrete to continuous: the broadening of number concepts in early modern England*. Dordrecht: Kluwer Academic Publishers.

Needham, Joseph. 1969. *The Great Titration: Science and Society East and West*. London: Allen & Unwin.

Nenci, Elio, eds. 2004. *Girolamo Cardano, De subtilitate*, Filosofia e scienza nell'età moderna 3/Testi inediti o rari 16. Milano: F. Angeli.

Olby, Robert, Geoffrey Cantor, John Christie, and M.J.S. Hodge, eds. 1990. *The Companion to the History of Modern Science*. London: Routledge.

Olmsted, John W. 1987. Jean Picard's 'Membership' in the Académie Royale des Sciences, 1666–1667: The Problem and Its Implications. In *Jean Picard*, ed. Picolet, 85–116. Paris: Éditions du Centre National de la Recherche Scientifique.

Ong, Walter J. 2004. *Ramus, Method, and the Decay of Dialogue: from the Art of Discourse to the Art of Reason*. Chicago: University of Chicago Press.

Osler, Margaret. 1998. Mixing Metaphors: Science and Religion or Natural Philosophy and Theology in Early Modern Science. *History of Science* 36: 91–113.

Ostwald, Jamel. 2006. Like Clockwork? Clausewitzian Friction and the Scientific Siege in the Age of Vauban. In *Instrumental in War*, ed. Walton, 85–117. Leiden: Brill.

Pantin, Isabelle. 2001. L' illustration des livres d' astronomie à la Renaissance: L' évolution d' une discipline à travers des images. In *Immagini per conoscere*, ed. F. Meroi and C. Pogliano, 3–41. Firenze: L.S. Olschki.

Parker, Geoffrey. 1988. *The Military Revolution. Military Innovation and the Rise of the West, 1500–1800*. Cambridge: Cambridge University Press.

Pauling, Linus. 1949, October. Chemistry and the World Today. An Invitation—and a Warning—to Private Industry to Come to the Aid of Basic Research. *Engineering and Science Monthly* 13(1): 5–8.

Pedersen, Olaf. 1978. The decline and fall of the *Theorica Planetarum*: Renaissance astronomy and the art of printing. *Studia Copernicana* 16: 157–185.

Perbellini, Gianni. 1994. *The Fortress of Nicosia, Prototype of European Renaissance Military Architecture*. Nicosia: Anastasios G. Leventis Foundation.

Perinbanayagam, R.S. 1990. How to do Self with Things. In *Beyond Goffman*, ed. S.H. Riggins, 315–340. Berlin: Mouton de Gruyter.

Petersen, Kurt. 2005. *Det Militære Målesystem: Kaliberstokken Og Dens Udvikling Fra 1540 Til 1850*. Lyngby/Denmark: Polyteknisk Forlag.

Pfaffenerger, Brian. 1999. Worlds in the Making: Technological Activities and the Construction of Intersubjective Meanings. In *Social Dynamics of Technology*, ed. C.R. Hoffman and M.-A. Dobres, 147–164. Washington, DC: Smithsonian Institution Press.

Picolet, Guy, eds. 1987. *Jean Picard et les Débuts de l'Astronomie de Précision au XVIIᵉ Siècle*. Paris: Édition du Centre National de la Recherche Scientifique.

Pisano, Raffaele, and Danilo Capecchi. 2015. *Tartaglia's Science of Weights and Mechanics in the Sixteenth Century*, History of Mechanism and Machine Science 28. Dordrecht: Springer.

Ploeg, Willem. 1934. *Constantijn Huygens en de Natuurwetenschappen*. Rotterdam: Nijgh & Van Ditmar.

Poovey, Mary. 1998. *A History of the Modern Fact: Problems of Knowledge in the Sciences of Wealth and Society*. Chicago: University of Chicago Press.

Popper, Nicholas. 2005. The English Polydaedali: How Gabriel Harvey Read Late Tudor London. *Journal of the History of Ideas* 66(3): 351–381.

Prager, Frank D., and Gustina Scaglia. 1970. *Brunelleschi: Studies of his Technology and Inventions*. Cambridge, MA: MIT Press.

———. 1972. *Mariano Taccola and his Book de Ingeneis*. Cambridge, MA: MIT Press.

Price, Derek J. de Solla. 1964. Automata and the Origins of Mechanism and the Mechanistic Philosophy. *Technology & Culture* 5(1): 9–23.

Pumfrey, Stephen. 2003. *Latitude and the Magnetic Earth*. Cambridge: Icon Books.

Pumfrey, Stephen, and Frances Dawbarn. 2004. Science and Patronage in England, 1570–1625: A Preliminary Study. *History of Science* 42: 157–160.

Pumfrey, Stephen, Paolo L. Rossi, and Maurice Slawinski, eds. 1991. *Science, Culture, and Popular Belief in Renaissance Europe*. Manchester: Manchester University Press.

Quinn, David Beers. 1955. *The Roanoke Voyages, 1584–1589: Documents to Illustrate the English Voyages to North America*. London: Hakluyt Society.

———. 1974. Thomas Harriot and the New World. In *Thomas Harriot*, ed. J.W. Shirley, 36–53. Oxford: Clarendon Press.

Rappaport, Steve. 1989. *Worlds within Worlds: Structures of Life in Sixteenth Century London*. Cambridge: Cambridge University Press.

Raven, Deiderick, Wolfgang Krohn, and Robert S. Cohen, eds. 2000. *Edgar Zilsel: The Social Origins of Modern Science*, Boston Studies in the Philosophy of Science 200. Dordrecht/Boston: Kluwer Academic Publishers.

Redknap, Mark, eds. 1997. *Artefacts from Wrecks: Dated Assemblages from the Late Middle Ages to the Industrial Revolution*. Oxford: Oxbow Books.

Reeves, Eileen. 2008. *Galileo's Glassworks: The Telescope and the Mirror*. Cambridge, MA: Harvard University Press.

Reid, Stephen, and Emma Annette Wilson, eds. 2011. *Ramus, Pedagogy, and the Liberal Arts: Ramism in Britain and the Wider World*. Farnham: Ashgate.

Renn, Jürgen, and Matteo Valleriani. 2001. Galileo and the Challenge of the Arsenal. *Nuncius* 16(2): 481–503.

Revelli, Paolo. 1929. *I codici Ambrosiani di contenuto geografico*. Milan: Luigi Alfieri.

Reynolds, Terry S. 1983. *Stronger Than a Hundred Men, a History of the Vertical Waterwheel*. Baltimore: Johns Hopkins University Press.

Ribe, Neal. 1997. Cartesian Optics and the Mastery of Nature. *Isis* 88: 42–61.

Richter, J.P., eds. 1970. *The Literary Works of Leonardo da Vinci*. London: Phaidon.

Rider, Robin Elaine. 1993. Early Modern Mathematics in Print. In *Non-Verbal Communication*, ed. R.G. Mazzolini, 91–113. Firenze: L.S. Olschki.

Rienitz, Joachim. 1993. 'Make Glasses to See the Moon Large': An Attempt to Outline the Early History of the Telescope. *Bulletin of the Scientific Instrument Society* 37: 7–9.

Riggins, Stephen Harold. 1990a. The Power of Things: the Role of Domestic Objects in the Presentation of Self. In *Beyond Goffman*, ed. S.H. Riggins, 347–349. Berlin: Mouton de Gruyter.

———, eds. 1990b. *Beyond Goffman: studies on communication, institution, and social interaction*. Berlin/New York: Mouton de Gruyter.

Roberts, Lissa. 2002. A World of Wonders, a World of One. In *Merchants & Marvels*, ed. P.H. Smith and P. Findlen, 399–411. New York: Routledge.

———, eds. 2011. *Centres and Cycles of Accumulation in and Around the Netherlands in the Early Modern Period*. Berlin: Lit.

Roberts, J., and A.G. Watson, eds. 1990. *John Dee's Library Catalogue*. London: Bibliographical Society.

Roberts, Lissa, Simon Schaffer, and Peter Dear, eds. 2007. *The Mindful Hand: Inquiry and invention from the late Renaissance to early industrialization*. Amsterdam: Koninkliijke Nederlandse Akademie van Wetenschappen.

Roche, Jon. Harriot, Thomas (c.1560–1621). In *DNB*. Online at http://www.oxforddnb.com/view/article/12379.

Ronan, Colin A. 1993. There Was an Elizabethan Telescope. *Bulletin of the Scientific Instrument Society* 37: 2–3.

Rose, Paul Lawrence. 1975. *The Italian Renaissance of Mathematics. Studies on Humanists and Mathematicians from Petrarch to Galileo*. Genève: Droz.

Rose, Susan. 2004. Mathematics and the Art of Navigation: The Advance of Scientific Seamanship in Elizabethan England. *Transactions of the Royal Historical Society* 14: 175–184.

Rose, Paul Lawrence, and Stillman Drake. 1971. The PseudoAristotelian Questions of Mechanics in Renaissance Culture. *Studies in the Renaissance* 18: 65–104.

Rossi, Paolo. 1970. *Philosophy, Technology and Arts in Early Modern Europe*. New York: Harper and Row.

Roth, Wolff-Michael. 1996. Where is the Context in Contextual Word Problems? *Cognition and Instruction* 14(4): 487–527.

Sahlins, Marshall. 1993. Goodbye to Tristes Tropes: Ethnography in the Context of Modern World History. *Journal of Modern History* 65: 1–25.

Sandman, Alison, and Eric H. Ash. 2004. Trading Expertise: Sebastian Cabot Between Spain and England. *Renaissance Quarterly* 57: 813–846.

Schemmel, Matthias. 2003. Thomas Harriot as an English Galileo: the Force of Shared Knowledge in Early Modern Mechanics. *Bulletin of the Society for Renaissance Studies* 21(1): 1–10.

———. 2008. *The English Galileo: Thomas Harriot's Work on Motion as an Example of Preclassical Mechanics*. Dordrecht: Springer.

Schiøler, T. 1973. *Roman and Islamic Waterwheels*. Odense: Odense University Press.

Schmidt, Wilhelm. 1902. Leonardo da Vinci und Heron von Alexandria. *Biblioteca Mathematica* 3(3): 180–187.

Schmitt, Charles B. 1967. Experimental Evidence for and Against a Void: The Sixteenth-Century Arguments. *Isis* 58: 352–366.

———. 1981. *Studies in Renaissance Philosophy and Science*. London: Variorum.

Schuster, John A. 1977. *Descartes and the Scientific Revolution: 1618–34: An Interpretation*. Ph.D. dissertation, Princeton University.

———. 1980. Descartes Opticien: The construction of the law of refraction . Descartes' Mathesis Universalis, 1619–28. In *Descartes: Philosophy, Mathematics and Physics*, ed. S. Gaukroger, 41–96. Sussex: Harvester Press.

———. 1990. The Scientific Revolution. In *Companion to the History of Modern Science*, ed. R.C. Olby et al., 217–242. London: Routledge.

———. 1995. Descartes *Agonistes* New Tales of Cartesian Mechanism. *Perspectives on Science* 3: 99–145.

———. 2000a. The construction of the law of refraction and the manufacture of its physical rationales, 1618–1629. In *Descartes' Natural Philosophy*, ed. S. Gaukroger, J.A. Schuster, and J. Sutton, 258–312. London: Routledge.

———. 2000b. Internalist and Externalist Historiographies of the Scientific Revolution. In *Encyclopedia of the Scientific Revolution*, ed. W. Appelbaum, 334–336. New York: Garland Publishing.

———. 2000c. Descartes Opticien: The construction of the law of refraction and the manufacture of its physical rationales, 1618–1629. In *Descartes' Natural Philosophy*, ed. S. Gaukroger et al., 258–312. London: Routledge.

———. 2002. L'Aristotelismo e le sue Alternative. In *La Rivoluzione Scientifica*, ed. D. Garber, 337–357. Rome: Istituto della Enciclopedia Italiana.

———. 2005. 'Waterworld': Descartes' Vortical Celestial Mechanics: A Gambit in the Natural Philosophical Contest of the Early Seventeenth Century. In *The Science of Nature*, ed. J.A. Schuster and P.R. Anstey, 35–79. Dordrecht: Springer.

———. 2012. Physico-mathematics and the Search for Causes in Descartes' Optics—1619–37. *Synthese* 185(3): 467–499.

———. 2013a. *Descartes-Agonistes Physico-Mathematics, Method & Corpuscular-Mechanism, 1618–33*, Studies in History and Philosophy of Science 27. New York: Springer.

———. 2013b. What was the relation of Baroque Culture to the Trajectory of Early Modern Natural Philosophy. In *Science in the Age of Baroque, Archives internationales d'histoire des idées*, ed. Ofer Gal and R. Chen-Morris, Vol. 208, 1–28. Dordrecht: Springer.

Schuster, John A., and Alan Taylor. 1996. Seized by the Spirit of Modern Science. *Metascience* 9: 9–26.

———. 1997. Blind Trust: The Gentlemanly Origins of Experimental Science. *Social Studies of Science* 27: 503–536.

Schuster, John A., and Graeme Watchirs. 1990. Natural Philosophy, Experiment and Discourse in the Eighteenth Century: Beyond the Kuhn/Bachelard Problematic. In *Experimental Inquiries*, ed. H.E. LeGrand, 1–48. Dordrecht: Kluwer Academic Publishers.

Schuster, John A., and Peter R. Anstey, eds. 2005. *The Science of Nature in the Seventeenth Century: Patterns of Change in Early Modern Natural Philosophy*. Dordrecht: Springer.

Shapin, Steven. 1982. History of Science and its Sociological Reconstructions. *History of Science* 20: 157–211.

———. 1988. The House of Experiment in Seventeenth-Century England. *Isis* 79: 373–404.

———. 1991. 'A Scholar and a Gentleman': The Problematic Identity of the Scientific Practitioner in Early Modern England. *History of Science* 29: 279–327.

———. 1992. Discipline and Bounding: The History and Sociology of Science As Seen Through the Externalism-Internalism Debate. *History of Science* 30: 333–369.

———. 1994. *A Social History of Truth: Civility and Science in Seventeenth-Century England*. Chicago: University of Chicago Press.

———. 1996. *The Scientific Revolution*. Chicago: University of Chicago Press.

Shapin, Steven, and Simon Schaffer. 1985. *Leviathan and the Air-Pump: Hobbes, Boyle, and the Experimental Life*. Princeton: Princeton University Press.

Shea, William R. 1991. *The Magic of Numbers and Motion. The Scientific Career of René Descartes*. Canton: Science History Publications, U.S.A..

Sherman, William H. 1995. *John Dee: The Politics of Reading and Writing in the English Renaissance*. Amherst: University of Massachusetts Press.

Shirley, John W., eds. 1974. *Thomas Harriot, Renaissance Scientist*. Oxford: Oxford University Press.

———. 1983. *Thomas Harriot: A Biography*. Oxford: Oxford University Press.

———. 1985. Science and Navigation in Renaissance England. In *Science and the Arts*, ed. J.W. Shirley and F.D. Hoeniger, 74–93. Washington, DC: Folger Shakespeare Library.

Shirley, John W., and F. David Hoeniger, eds. 1985. *Science and the Arts in the Renaissance*. Washington, DC: Folger Shakespeare Library.

Shumaker, Wayne, ed. and trans. 1978. *John Dee on Astronomy, Propaedeumate Aphoristica (1558 and 1568), Latin and English*. Berkeley: University of California Press.

Skinner, Quentin. 1969. Meaning and Understanding in the History of Ideas. *History and Theory* 8: 3–53.

Smith, Ole L. 1973. On Some Manuscripts of Heron, Pneumatica. *Scriptorium* 27(1): 96–101.

Smith, A. Mark. 1976. Galileo's Theory of Indivisibles: Revolution or Compromise? *Journal of the History of Ideas* 37: 571–588.

———. 1981. Getting the Big Picture in Perspectivist Optics. *Isis* 72: 568–589.

Smith, Robert D., eds. 1989. *British Naval Armaments*, Royal Armouries conference proceedings 1. London: Royal Armouries.

Smith, Pamela H. 1994. *The Business of Alchemy: Science and Culture in the Holy Roman Empire*. Princeton: Princeton University Press.

———. 2004. *The Body of the Artisan. Art and Experience in the Scientific Revolution*. Chicago: University of Chicago Press.

Smith, Pamela H., and Paula Findlen, eds. 2002. *Merchants & Marvels: Commerce, Science and Art in Early Modern Europe*. New York: Routledge.

Smuts, R. Malcolm. 1987. *Court Culture and the Origins of a Royalist Tradition in Early Stuart England*. Philadelphia: University of Pennsylvania Press.

Snelders, H.A.M.. 1980. Alkmaarse natuurwetenschappers uit de 16de en 17de eeuw." 101–122. In *Van Spaanse beleg tot Bataafse tijd. Alkmaars stedelijk leven in de 17de en 18de eeuw*. Alkmaarse historische reeks 4. Zutphen.

Solomon, Julie Robin, and Catherine Gimelli Martin, eds. 2005. *Francis Bacon and the refiguring of early modern thought: essays to commemorate the advancement of learning (1605–2005)*. Aldershot: Ashgate.

Star, Susan Leigh. 1989. The Structure of Ill-Structured Solutions: Boundary Objects and Heterogeneous Distributed Problem Solving. In *Distributed Artificial Intelligence*, ed. Gasser and Huhns, Vol. II, 37–54. London: Pitman.

Star, Susan Leigh, and James Griesemer. 1989. Institutional Ecology, 'Translations' and Boundary Objects: Amateurs and Professionals in Berkeley's Museum of Vertebrate Zoology, 1907–39. *Social Studies of Science* 19(3): 387–420.

Steele, Brett D. 1994. Muskets and Pendulums: Benjamin Robins, Leonhard Euler, and the Ballistics Revolution. *Technology & Culture* 35: 348–382.

Stern, Virginia F. 1979. *Gabriel Harvey: His Life, Marginalia, and Library*. Oxford: Clarendon Press.

Stewart, Larry. 1992. *The Rise of Public Science: Rhetoric, Technology, and Natural Philosophy in Newtonian Britain*. Cambridge: Cambridge University Press.

Stewart, Richard W. 1996. *The English Ordnance Office, 1585–1625. A Case Study in Bureaucracy*. Woodbridge: Boydell.

Stone, Lawrence. 1964. The Educational Revolution in England, 1560–1640. *Past & Present* 28: 41–80.

Strong, Roy. 1986. *Henry, Prince of Wales and England's Lost Renaissance*. London: Thames and Hudson.

Szulakowska, Urszula. 2000. *The Alchemy of Light: Geometry and Optics in Late Renaissance Alchemical Illustration*. Leiden: Brill.

Taverne, E. 1978. *In 't land van belofte: in de nieue stadt. Ideaal en werkelijkheid van de stadsuitleg in de Republiek. 1580–1680*. Schwartz: Maarssen.

Taylor, E.G.R. 1928. William Bourne: A Chapter in Tudor Geography. *The Geographical Journal* 72(4): 329–339.

———. 1930. *Tudor Geography 1485–1583*. London: Methuen.

———. 1934. *Late Tudor and Early Stuart Geography 1583–1650*. London: Methuen.

———. 1954. *The Mathematical Practitioners of Tudor and Stuart England*. Cambridge: Cambridge University Press.

———. 1958. *The Haven-Finding Art: A History of Navigation froSm Odysseus to Captain Cook*. London: Hollis & Carter.

———. 1963. *A Regiment for the Sea and other Writings on Navigation by William Bourne*. Cambridge: Cambridge University Press.

———. 1966. *The Mathematical Practitioners of Hanoverian England, 1714–1840*. Cambridge: Cambridge University Press.

Taylor, Katie. 2013. Reconstructing Vernacular Mathematics: The Case of Thomas Hood's Sector. *Early Science and Medicine* 18(1–2): 153–179.

Terpstra, Harke, and Chriet Titulaer. 1981. *Friesche Sterrekonst. Geschiedenis van de Friese sterrenkunde en aanverwante wetenschappen door de eeuwen heen*. Franeker: Wever.

Thorndike, Lynn. 1923–58. *History of Magic and Experimental Science*. Vol. 8 vols. New York: The Macmillan Company.

Thorpe, William Arnold. 1949. *English Glass*. London: Black.

Todd, H.J. 1812. *A Catalogue of the Archepiscopal Manuscripts in the Library at Lambeth Palace*. London: Lambeth Palace Library.

Turner, G.L'E. Bourne, William (*c.* 1535–1582). In *DNB*. Oxford University Press, 2004. http://www.oxforddnb.com/view/article/3011. Accessed 8 June 2016.

Turner, A.J. 1973. Mathematical Instruments and the Education of Gentlemen. *Annals of Science* 30: 51–88.

Turner, G.L'E. 1995. The Instruments Makers of Elizabethan England. *Sartoniana* [Ghent], 8: 19–31.

———. 2000. *Elizabethan Instrument Makers: the Origins of the London Trade in Precision Instrument Making*. Oxford: Oxford University Press.

UNC University Library. 2006. *Picturing the New World. The Hand-Coloured De Bry Engravings of 1590*. University of North Carolina at Chapel Hill. Online at http://www.lib.unc.edu/dc/debry.

Valleriani, Matteo. 2009. The Transformation of Aristotle's Mechanical Questions: A Bridge Between the Italian Renaissance Architects and Galileo's First New Science. *Annals of Science* 66(2): 183–208.

———. 2010. *Galileo, Engineer*. Dordrecht: Springer.

Valleriani, Matteo, Lindy Divarci, and Anna Siebold. 2013. *Metallurgy, Ballistics, and Epistemic Instruments: the Nova Scientia of Nicolò Tartaglia*. Berlin: Edition Open Access.

Van Berkel, Klass. 1985. Het onderwijs in de wiskunde in Franeker in vergelijkend perspectief. *It Beaken* 47: 220–222.

———. 1999. The Legacy of Stevin. A Chronological Narrative. In *History of Science in the Netherlands*, ed. K. van Berkel et al., 16–20. Leiden: Brill.

Van Berkel, Klass, Albert van Helden, and Lodewijk Palm, eds. 1999. *A History of Science in the Netherlands. Survey, Themes and Reference*. Brill: Leiden.

Van Bunge, W., et al., eds. 2003. *The Dictionary of Seventeenth and Eighteenth-Century Dutch Philosophers*. Vol. 2. vols. Thoemmes: Bristol.

Van den Heuvel, Ch.. 2000. Wisconstighe Ghedachtenissen. Maurits over de kunsten en wetenschappen in het werk van Stevin. In *Marits,* eds. Zandvliet and Eyffinger, 113–116.

Van Dyck, M. 2006. Gravitating toward Stability: Guidobaldo's Aristotelian-Archimedean Synthesis. *History of Science* 44: 373–407.

Van Helden, Albert. 1977. *The Invention of the Telescope*, Transactions of the American Philosophical Society 67, part 4. Philadelphia: American Philosophical Society.

Van Schooten, F.. 1974–86. Calcul de Mons. Des Cartes. 10. In *Oeuvres,* ed. Descartes, 659–680.

Van Winter, P.J. 1988. *Hoger beroepsonderwijs avant-la-lettre: Bemoeiingen met de vorming avn landmeters en ingenieurs bij de Nederlandse universiteiten van de 17ᵉen 18ᵉeeuw*, Verhandelingen der Koninklijke Nederlandse Akademie van Wetenschappen, Afd. Letterkunde, Nieuwe Reeks 137. Amsterdam: Noord-Hollandsche Uitg. Mij.

Vanden Broecke, Steven. 2001. Dee, Mercator, and Louvain Instrument Making: An Undescribed Astrological Disc by Gerard Mercator (1551). *Annals of Science* 58: 226–229.

Vermij, R. 2003. The Formation of the Newtonian Philosophy: The Case of the Amsterdam Mathematical Amateurs. *The British Journal for the History of Science* 36: 183–200.

Wallis, Helen M. 1951. The First English Globe: A Recent Discovery. *The Geographical Journal* 117(3): 275–290.

———. 1952. The Molyneux Globes. *British Museum Quarterly* 16: 89–90.

———. 1999. 'Opera Mundi': Emery Molyneux, Jodocus Hondius and the First English Globes. In *Theatrum Orbis Librorum,* eds. Croiset van Uchelen, et al., 94–104.

Walton, Steven A. 1999. *Thomas Harriot's Ballistics and English Renaissance Warfare.* Durham Thomas Harriot Seminar Occasional Paper 30. Durham.

———. 2003. The Bishopsgate Artillery Garden and the First English Ordnance School. *Journal of the Ordnance Society* 15: 41–51.

———., ed. 2005a. *Instrumental in War: Science, Research, and Instruments Between Knowledge and the World*, History of Warfare 28. Leiden: Brill.

———. 2005b. Mathematical Instruments and the Creation of the Scientific Military Gentleman. In *Instrumental in War*, ed. Walton, 17–46. Leiden: Brill.

———. 2006. The Tower Gunners and the Artillery Company in the Artillery Garden before 1630. *Journal of the Ordnance Society* 18: 53–66.

———. 2009. reCOGnition: Medieval Gearing from Vitruvius to Print. *AVISTA Forum Journal* 19(1/2): 28–41.

———. 2010. State Building Through Building for the State: Domestic and Foreign Expertise in Tudor Fortifications. In *Expertise and the Early Modern State. Osiris* 25, ed. Eric Ash, 66–84.

———. 2013. Proto-Scientific Revolution or Cookbook Science? Early Gunnery Manuals in the Craft Treatise Tradition. In *Craft Treatises*, ed. Cordoba, 221–236. Turnhout: Brepols.

Warner, Deborah. 1990. What is a Scientific Instrument, When Did it Become One, and Why? *British Journal for the History of Science* 23: 83–93.

Waters, David W. 1958. *The Art of Navigation in England in Elizabethan and Early Stuart Times.* New Haven: Yale University Press.

Weaver, David S. 1995. The English Gunner's Caliper. *Arms Collecting* 33: 111–125.

Webb, Henry J. 1965. *Elizabethan Military Science: The Books and the Practice*. Madison: University of Wisconsin Press.

Webster, Charles. 2002. *The Great Instauration: Science, Medicine and Reform 1626–1660.* 2d ed. 1976. Oxford: Oxford University Press.

Weeks, Andrew. 1997. *Paracelsus. Speculative Theory and the Crisis of the Early Reformation.* Albany: State University of New York Press.

Weinberger, Jerry. 2005. Francis Bacon and the unity of knowledge: reason and revelation. In *Francis Bacon*, ed. Solomon and Martin, 109–127. Aldershot: Ashgate Publishing.

Westra, Frans. 1992. *Nederlandse ingenieurs en de fortificatiewerken in het eerste tijdperk van de Tachtigjarige Oorlog, 1573–1604*. Alphen aan de Rijn: Canaletto.

Whitaker, Ewen A. 1993. The Digges-Bourne Telescope – An Alternative Possibility. *Journal of the British Astronomical Association* 103: 310–312.

White, Lynn Jr. 1967. Jacopo Acontio as an Engineer. *American Historical Review* 72(2): 425–444.

———. 1978. *Medieval Religion and Technology: Collected Essays*. Berkeley: University of California Press.

Wolfe, Jessica. 2004. *Humanism, Machinery and Renaissance Literature*. Cambridge: Cambridge University Press.

Yachnin, Paul, and Bronwen Wilson, eds. 2009. *Making Publics in Early Modern Europe: People, Things, Forms of Knowledge*. London: Taylor & Francis.

Yates, Frances. 1972. *The Rosecrucian Enlightenment.* London: Routledge/Kegan Paul.

Zandvliet, Kees, and Arthur Eyffinger, eds. 2000. *Maurits, Prins van Oranje*. Zwolle: Waanders.

Zilsel, Edgar. 1940. Copernicus and Mechanics. *Journal of the History of Ideas* 1: 113–118.

———. 1941. The Origins of Gilbert's Scientific Method. *Journal of the History of Ideas* 2: 1.

———. 1942a. The Genesis of the Concept of Physical Law. *Philosophical Review* 51: 245–279.

———. 1942b. The Sociological Roots of Science. *American Journal of Sociology* 47: 552–555.

———. 1945. The Genesis of the Concept of Scientific Progress. *Journal of the History of Ideas* 6: 325–349.

———. 1957. *Roots of Scientific Thought*, ed. Philip P. Wiener and Aaron Noland. New York: Basic Books.

———. 2000. *The Social Origins of Modern Science*, Boston Studies in the Philosophy of Science 200, ed. Diederick Raven et al. Dordrecht: Kluwer Academic Publishers.

Zuidervaart, Huib J. 1995. *Speculatie, Wetenschap en Vernuft. Fysica en astronomie volgens Wytze Foppes Dongjuma (1707–1778), instrumentmaker te Leeuwarden*. Leeuwarden: Fryske Akad..

———. 2004. Reflecting 'Popular Culture': The Introduction, Diffusion, and Construction of the Reflecting Telescope in the Netherlands. *Annals of Science* 61: 407–452.

———. 2012. The 'Invisible Technician' Made Visible: Telescope making in the Seventeenth and early Eighteenth-century Dutch Republic. In *From Earth-Bound to Satellite*, ed. Morrison-Low et al., 41–102. Leiden: Brill.

The manufacturer's authorised representative in the EU is Springer Nature Customer Service Centre GmbH, Europaplatz 3, 69115 Heidelberg, Germany. If you have any concerns regarding our products, please contact ProductSafety@springernature.com

Printed and bound by CPI Group (UK) Ltd, Croydon, CR0 4YY
01/10/2026
02225870-0001